WMU Studies in Maritime Affairs

Volume 10

WMU Studies in Maritime Affairs was launched in 2013 to inspire scholars in the field of maritime affairs to contribute to the advancement of knowledge through publications of the highest order of excellence. With this book series, the World Maritime University aims to lead an expansion of scholarly pursuits, particularly in the areas of maritime safety and environmental administration, maritime law and policy, maritime education and training, maritime energy management, ocean sustainability and governance, and shipping and port management and logistics.

Please contact the Series Editors (joma@wmu.se) for more information.

More information about this series at http://www.springer.com/series/11556

Byoung-Wook Ko • Dong-Wook Song
Editors

New Maritime Business

Uncertainty, Sustainability, Technology and Big Data

Springer

Editors
Byoung-Wook Ko
Korea Maritime Institute
Busan, Republic of Korea

Dong-Wook Song
World Maritime University
Malmo, Sweden

ISSN 2196-8772 ISSN 2196-8780 (electronic)
WMU Studies in Maritime Affairs
ISBN 978-3-030-78956-5 ISBN 978-3-030-78957-2 (eBook)
https://doi.org/10.1007/978-3-030-78957-2

This Springer imprint is published by the registered company Springer Nature Switzerland AG.
The registered company address is: Gewerbestrasse 11, 6330 Cham, Switzerland

Foreword

This publication is the outcome of the seminar jointly held between the World Maritime University (WMU) and the Korea Maritime Institute (KMI). It is a product of the successful collaboration over the past few years between the two institutions. Our relationship with KMI and the Republic of Korea goes much beyond this event. To date, there are over 100 WMU graduates from the Republic of Korea. They have been making an important contribution to the country including leading various maritime organizations not only in the country but also globally. The Secretary-General of IMO, Mr. Kitack Lim is an outstanding example of the contribution he has made to the country but also the impact he is making in his current functions at the IMO. The President of KMI, Dr. Young-Tae Chang, is also a shining example of the impact of WMU with the various roles he has held including his current position. We also take pride in having one member of our faculty members, Professor Daniel Moon, who is currently serving as the Minister of Oceans and Fisheries for the Republic of Korea. We are very proud of all of them.

This seminar falls within the framework of the MOU signed between WMU and KMI in June 2016, which provides for the holding of a joint symposium, on an annual basis, to develop the modalities for mutual cooperation and collaboration and high-quality research. These seminars enable us to address critical issues facing the maritime industry and to identify potential options and opportunities available. The 2020 seminar was unique as it was conducted virtually, which demonstrates the strong motivation for the two organizations to collaborate and develop maritime business and strategies, especially during these difficult times. A working group comprising of faculty members at WMU and researchers at KMI addressed several topics on 'Container Shipping Business and Maritime 4.0: Policy and Strategy'.

To address some of the fundamental questions from the maritime business perspective, three different sub-sessions were held. They covered:

- the recent developments in the container shipping market;
- several recent changes relating to environmental regulations and their impact on the shipping ecosystem;

- Maritime 4.0 and the development of new business models for the shipping industry.

The seminar's output has an impact not only on national and international trade but also on society and the maritime industry as a whole. An invitation was sent to several well-known experts in the industry and in academia. They have shared their experience and knowledge and sought to bridge the gap between theory and practice. Following the seminar, each presenter was invited to contribute to the 'WMU Studies in Maritime Affairs' book series. This will help the industry design policies for the maritime industry and support academic research.

It is my hope that the book will provide useful insights on the state of the industry today, the lessons learnt as well as the outlook for the future. I wish to take this opportunity to personally thank KMI and all participants for their engagement, despite the difficult times, making this seminar a success. I look forward to the next seminar in due course.

World Maritime University, Malmo, Sweden — Cleopatra Doumbia-Henry

Foreword

The Korea Maritime Institute (KMI) is a government-sponsored research institute based in the Republic of Korea. KMI is committed to becoming a 'Global Center of Excellence in Maritime and Fisheries Research'. Since 2016, KMI and the World Maritime University (WMU) have jointly held the 'KMI-WMU Seminar' on an annual basis, which has played an important role by creating a platform for cooperation in the field of maritime research.

Our global maritime society is facing a number of serious challenges posed by climate change, safety issues, and shipping innovations via the 4th Industrial Revolutions. That is, we must overcome climate change and ensure citizen's safety by using technologies from the 4th Industrial Revolution. In addition, the outbreak and worldwide spread of COVID-19 have called for the innovation of our cooperation capacity. In this background, KMI and WMU held the 4th KMI-WMU Seminar on 12–13 November 2020 in a hybrid form, combining online and offline sessions in Seoul, Korea. The seminar entitled 'Container Shipping Business and Maritime 4.0: Policy and Strategy' consisted of two sessions: the Industry Session and the Academic Session.

This book, adequately named 'New Maritime Business: Uncertainty, Sustainability, Technology, and Big Data', is a collection of eight presentation-based chapters and three additionally invited chapters. All the contributors are active researchers in their fields who have written excellent articles on the subject. Part I focuses on uncertainty, while Part II is on climate change with Parts III and IV covering the 4th Industrial Revolution and its related topics. In this foreword, I express my deep gratitude to those contributors.

I believe that this book is an important first step towards even closer cooperation among the KMI, WMU, and global scholars. In order for our maritime academic communities to lead the maritime industries, we, including KMI and WMU, should continue to conduct in-depth research with high practical applicability and excellence in quality. These kinds of multidisciplinary efforts that enabled the construction of this book should be further expanded. In this line of efforts, KMI and WMU will continually endeavour to make a ground for such mutually beneficial research environment in a form of online, offline seminars and/or publications of books.

Finally, I hope that this book will help in improving our capability to cope with future uncertainties and then enhancing the performances of its reader's organizations. Furthermore, this book is expected to not only provide a useful compass for navigating rapidly changing markets but also be utilized as a scientific reference for readers' further investigations.

Korea Maritime Institute, Busan, South Korea

Young-Tae Chang

Preface

Two premier research and academic institutes in the maritime affairs, the Korea Maritime Institute (KMI) and World Maritime University (WMU) convened a seminar under the title 'Container Shipping Business and Maritime 4.0: Policy and Strategy' in November 2020. Unlike the previous three seminars jointly held, the 4th Seminar was conducted in a hybrid manner, having the two sessions: Industry Session and Academic Session. This book you are holding on is a collection of eight presentation-based chapters from the Academic Session and three invitation-based ones.

Back to the preparation stage in the early 2020, Young-Tae Chang, KMI President, suggested a few agendas that were to be discussed for the seminar. While having an initial brainstorming session among the KMI team members, composed of Hee-Sung Yun, Han-Seon Park, and Byoung-Wook Ko, we summarized the following three themes for the academic session: recent developments from COVID-19, environmental regulation and responses of shipping ecosystem, and Maritime 4.0 and new business opportunities. With the endorsement of Cleopatra Doumbia-Henry, WMU President, Dong-Wook Song invited the prominent scholars in those suggested fields as both the speakers of the seminar and contributors of this book. Having successfully finished the seminar, despite a number of technical challenges, the two editors started working for the book by arranging a cross-reviewing process of the submitted chapters as well as by making further invitations to be contributors in the topics that could not be covered at the seminar due to time- and space-constraints. Furthermore, the third theme of the seminar, Maritime 4.0, was divided into the two sub-themes, i.e., digitalization and big data, and technology and new business model.

It is hoped that our deliberate choice of the themes for this book and the chapters contained herein would give stakeholders in the industry, such as policymakers and practical professionals, and more importantly scientific communities in the academic discipline, an insight into the nature of ever-changing and greater uncertainty facing in the industry now and future. We do also believe that the chapters to be presented in the book initiate and even provoke an innovative way of doing businesses and implementing policies, so that our beloved maritime industry becomes better informed, prepared, and greatly contributing to human lives.

In addition to the above-mentioned people, we would like to express our deep gratitude to H.E. Daniel Seong-Hyeok Moon, the Minster of Oceans and Fisheries, Republic of Korea, who has been a full supporter for the seminar from the beginning. We are indebted to the following two companies—HMM and Busan Port Authority—for their great help in preparing the seminar. Most importantly, all the contributors of the chapters are to be thankfully acknowledged; without their professionally geared quality and punctual commitment from the beginning to now, this book would have not been appeared as it is now.

The publication of this book has been possible in a team-work spirit between KMI and WMU. Personally, however, Byoung-Wook Ko played his role as a co-editor being a representative of KMI on his passion as an economist towards a better understanding of the industry. He treats this publication journey as an invaluable experience and opportunity for making himself a contributor to the 'betterness' of maritime industry as well as the advancement of academic communities in the field.

Having made a due effort to present all the chapters in a professional manner, however, we do also acknowledge that there might be an error—hopefully, not many—noted throughout the pages to be followed. Please freely let us know about it, so that we (including you) could *collectively* make our discipline a practically sensible and academically sound field.

Enjoy reading the chapters.

Busan, Republic of Korea
Malmo, Sweden
15 February 2021

Byoung-Wook Ko
Dong-Wook Song

Contents

Introduction 1
Dong-Wook Song

Part I Recent Market Development

Volatility and Uncertainty in Container Shipping Market 11
Jasmine Siu Lee Lam, Qingyao Li, and Shuyi Pu

Containerships and Black Swans: The Impact of Disruptive Events on the Industry 33
Max Johns

COVID-19: Challenges and Future Responses in International Logistics 55
Sung-Jun Park and Eun-Soo Kim

Part II Decarbonization and Green Finance

Container Shipping Decarbonization Pathways 75
Pierre Cariou and Elizabeth Lindstad

Green Shipping Finance: Existing Initiatives and the Road Ahead 95
Manolis G. Kavussanos and Dimitris A. Tsouknidis

Part III Digitalization and Maritime Big Data

Maritime Digital Business Models: The Case of D/S Norden, 2015–2020 113
Martin Jes Iversen

Platform Revolution in Container Shipping Markets: An Economics Perspective 127
Byoung-Wook Ko

Liner Services in Asia's Major Container Ports: Implications for Busan 147
Satya Sahoo

Part IV Technology and New Business Model

Research Developments and Debates Regarding Maritime Autonomous Surface Ship: Status, Challenges and Perspectives 175
Tae-eun Kim and Jens-Uwe Schröder-Hinrichs

New Business Models for Shipping: Innovation in the Netherlands 199
Albert W. Veenstra

Editors and Contributors

About the Editors

Byoung-Wook Ko is Director of Shipping Policy Research Department at the Korea Maritime Institute. His main research interests include an econometrics-based analysis on markets, industrial policy and strategy, seaborne trade in shipping and shipbuilding industries, and shipping finance. He won the STX Prize Award in 2011 with one of his papers: 'A Mixed-Regime Model for Dry Bulk Freight Market' (2010) selected by the editorial board of *Asian Journal of Shipping and Logistics*. He has worked for the Korean government on shipping industrial policy design and evaluation and also been providing consulting services to market participants in Korean shipping ecosystems such as shipping firms, associations, financial institutes, shippers, and so forth. He is currently serving as a member of editorial board of *KMI International Journal of Maritime Affairs and Fisheries*. Among his research works, the paper, entitled 'The Nationality Effect from Price Discrimination in the Container Shipping Market and the Implications for Development Economics: Treating Hanjin Shipping's Collapse as a Natural Experiment' (2020) published in *Journal of Korea Trade*, is well received and regarded as an important contribution to the global governance design in container shipping markets. In addition, another paper of his: 'Dynamic Patterns of Dry Bulk Freight Spot Rates through the Lens of a Time-Varying Coefficient Model' (2018) appeared in *Transportation Research Part A*, is considered a path-breaking article as for its econometric treatment of shipping sectors' empirics. While keeping the above research interests under his research umbrella, he has recently made an endeavour towards understanding knowledge components in shipping ecosystems as an emerging research topic with practicality in mind.

Dong-Wook Song is Professor of Maritime Economics and Logistics and Korean Chair in Maritime Affairs at the World Maritime University, Sweden. Prior to the current post, he worked at universities in the UK and Hong Kong. He holds a BA (Hons) with First Class in Shipping Management from Korea Maritime University, an MSc in International Shipping and Logistics, and a PhD in Maritime Economics and Logistics from the University of Plymouth, UK. He also attended an Industrial Economics Course at the London School of Economics. He is currently an Editor-in-Chief of *WMU Journal of Maritime Affairs* and sits on editorial boards at highly regarded transport and logistics journals. He has worked as a consultant for a number of private and public organizations including the UN, OECD, Asian Development Bank, and World Bank. He served the International Association of Maritime Economists, a premier academic association in the field of maritime transport and logistics, as an elected Council Member for 12 years (2004–2010; 2012–2018), being among the longest-serving Council Members for the Association. One of his co-authored papers in 2006 is recently enlisted by Google Scholar (2017) as a classic paper in the entire field of transportation. In addition, an article in 'Transport Reviews' (2013) shows that his works in port research are highly ranked—among the top listed over the last three decades (1980–2009). Another article in 'Maritime Policy and Management' (2013) lists his two papers, published in 2003 and 2002, that are among the top 30 most cited papers over the whole 40-year history of the journal (1973–2013). His prolific but impact-bearing-refereed publications are a product of his keen interest in managerial and strategic aspects of global shipping, port, and logistics. See Citation Index for more of his research works.

Contributors

Pierre Cariou is Senior Professor at Kedge Business School in Bordeaux, France. He holds a PhD in Applied Economics from the University of Nantes. Prior to the current position, he worked as Professor as the World Maritime University in Sweden (2004–2010) and Associate Professor in Economics at the Faculty of Economics of Nantes (2001–2004). He contributed to several reports for private (CMA CGM, Natixis Bank, Casino Group…) and public (OMI, OECD, French Parliament, Port Authorities of Nantes and Marseille) organizations on shipping/port economics and published more than 60 academic papers. He is a member of KEDGE Supply Chain Centre of Excellence (CESIT) and of the Khüne Logistics University/ Hapag-Lloyd of Global Logistics Research Centre.

Martin Jes Iversen is Vice Dean of International Education and Associate Professor at Department for Strategy and Innovation and Copenhagen Business School, Denmark. His research is focused on the maritime sector with recent publications about the development of Danish shipping in the twenty-first century (Palgrave MacMillan, 2020). Prior to the current position as Vice Dean Martin Jes Iversen was Program Director of the BSc in International Shipping and Trade program at CBS. He has been a visiting scholar at UC Berkeley and Stockholm University. He has recently been appointed MPA Professor in Maritime Economics at Singapore Management University and he will be teaching and studying at SMU in the spring of 2022.

Max Johns is Professor and Vice President at Hamburg School of Business Administration and Head of its Maritime Business School in Hamburg, Germany, and is a Guest Professor at World Maritime University. He holds a BA and MA from Université Paul Valèry in Montpellier, France, an MBA from KU Leuven, Belgium and a PhD from Eberhard Karls University in Tübingen, Germany. He serves as senior advisor to the German shipowners and as such is Chair of the shipowners' group at ILO and has been a chief negotiator for ILO minimum wages in 2016 and 2018. Prior to the current post, Dr. Johns was Managing Director of the German Shipowners Association (VDR) and had an extensive career in the media industry where he was CEO of ZDF.newmedia and SVP of New Media at public broadcaster ZDF in Germany.

Manolis Kavussanos is Professor at Athens University of Economics and Business, Greece, founder and Director of the MSc in International Shipping, Finance and Management and the Laboratory with the corresponding name. He has been Director of the MSc and PhD programs in Accounting and Finance at AUEB; co-launched and directed the MSc in Trade, Logistics and Finance at Cass Business School, London; Council member of IAME; BoD member of Institute of History of Commercial Shipping; editorial board member of major international scientific journals; award committees' member for prestigious academic/professional prizes/ scholarships, including Onassis foundation; received teaching awards at Cass and AUEB, IAME research-paper awards, Propontis foundation post-doc academic scholarship; visiting professor of finance and shipping in universities in UK, France, Netherlands, Belgium, Italy, Singapore, China, Hong Kong, Greece and Cyprus. Since 1992 he has worked in developing the area of risk analysis and management in shipping and is co-author of four books in this area and in shipping finance.

Eun-Soo Kim is Director of International Logistics Investment Analysis and Support Centre, Shipping and Logistics Research Division at the Korea Maritime Institute. Prior to the current position, he worked as a Representative of KMI China Research Centre, in Shanghai, China (2010–2014). He holds a BA in Agricultural Civil Engineering from Seoul National University, an MSc in Management Science of Business Administration from Yonsei University, Korea and a PhD in Transport Planning and Management from Shanghai Maritime University, China. He has been conducting research projects on agendas and government planning in various maritime logistics related fields including container shipping, seaport strategies, international logistics and related supply chain management in KMI since 2004.

Tae-eun Kim is Associate Professor of Maritime Safety at the University of South-Eastern Norway. She has an MSc in maritime management from USN and a joint PhD in nautical operations from four Norwegian national universities offering maritime education. Her research focuses on maritime safety and human factors, including accident causation modelling, safety leadership, and management. She is currently working as the co-leader for Focus Area 2 of the Norwegian Centre of Excellence in Maritime Simulator Training and Assessment (COAST) and the deputy leader for the research group in Maritime Safety, Digitalization, Leadership and Innovation (MASDLI) at USN.

Jasmine Siu Lee Lam is Associate Professor and Director of the Maritime Energy and Sustainable Development Centre of Excellence at Nanyang Technological University, Singapore. She is an active scholar having completed over 50 R&D projects, 270 publications in leading international journals, books and conferences. She serves as the editor/board member of 10 international journals, including the flagship maritime journal Maritime Policy and Management. She is the recipient of many awards, including the National Day Honours Commendation Medal awarded by the President of Singapore, 8 times Best Paper Awards, Eagle Prize, and Erasmus Mundus Scholars Award. She received her PhD from University of Antwerp, Belgium.

Qingyao Li graduated from Nanyang Technological University, Singapore with a BSc in Maritime Studies with honours (highest distinction) in 2019. She is currently pursuing a PhD degree in the same university. She is also a Project Officer at the School of Civil and Environmental Engineering, NTU. Her interest in research is related to port technology in port development, such as port automation and digitalisation. She has contributed and participated in various events and has taken charge of projects in her field of research.

Elizabeth Lindstad is Chief Scientist in SINTEF Ocean, Norway. She holds a MSc in Industrial Economics from the Norwegian Institute of Technology and a PhD in Naval Architecture from the Norwegian University of Science and Technology. She has long experience in developing and managing research and development projects with focus on sustainable shipping solutions, new innovative vessel designs and supply chains. She has been co-author, of EU & IMO policy studies. Member of ESSF the European Sustainable Shipping Forum, consisting of representatives from the European shipping sector and national representatives since it was established in 2013.

Sung-Jun Park is Research Fellow working at the International Logistics Investment Analysis and Support Centre, Korea Maritime Institute. His research is focused on bilateral and multilateral cooperation of inter-modal transport and logistics sector in Northeast Asia. Prior to the current position, he worked as the Director of Korea-Russia Research Centre at the Maritime State University located in Vladivostok, Russia (2018–2019), conducting joint research projects for nurturing governmental joint efforts for enhancing international logistics interconnectivity in the Russian Far East region. He holds a BA in political science from Korea University, obtained his MD from Monterey Institute of International Studies in USA, and holds a PhD in North Korea Studies from Korea University, Republic of Korea.

Shuyi Pu received an MSc in Supply Chain Management from National University of Singapore in 2015. She is currently pursuing a PhD in Maritime Studies at Nanyang Technological University, Singapore, while working as a research associate at the same university. Her current research interests include blockchain, digitalisation, emission accounting and green shipping in the maritime industry. She has published papers in highly regarded maritime journals such as Maritime Policy and Management and prestigious conferences such as IEEM.

Satya Sahoo is Assistant Professor of Shipping Management and Finance at the World Maritime University, Sweden. His research interests include big data analytics in maritime, shipping finance, and maritime asset management. He received a PhD in Finance from Henley Business School, University of Reading, UK. The ICMA Centre scholarship sponsored his doctoral research. He holds an MSc in Maritime Affairs specializing in Shipping Management and Logistics from the World Maritime University, Sweden, and a BEng in Marine Engineering from the Indian Maritime University: Kolkata Campus, India. He did previously work at the Fleet Management Ltd. Hong Kong as a marine engineer before embarking his academic career.

Jens-Uwe Schröder-Hinrichs is Professor and Vice-President (Academic Affairs) of the World Maritime University. He is an internationally recognized expert in the field of maritime administration, risk and safety, accident causation modelling, implementation and enforcement of IMO instruments, and capacity building. He oversees and directs the development of the academic agenda of WMU to strengthen the University's academic profile and status as the global centre of excellence in maritime education, research, and capacity building. With his extensive academic experience and seafaring background, he maintains active links and collaborations with the maritime industry, MET institutions, as well as the seafaring communities and policy-making authorities across the globe.

Dimitris A. Tsouknidis is Associate Professor of Finance with emphasis in Shipping at Athens University of Economics and Business, Greece. He has been a full-time faculty member at the University of Piraeus, Cyprus University of Technology and University of Bradford; also Visiting Faculty member at the Athens University of Economics and Business, Hellenic Open University, Greece, and Henley Business School, University of Reading, UK. He has published in highly ranked refereed journals, co-authored books and book chapters, and serves as Associate Editor of the *Maritime Policy and Management* journal. He has received a best paper award at the International Association of Maritime Economists (IAME) Conference and a teaching excellence award from the Athens University of Economics and Business.

Albert W. Veenstra is Professor of Trade and Logistics at the Rotterdam School of Management, and Scientific Director of the Dutch Institute for Advanced Logistics (Dinalog). He received a PhD from the Erasmus School of Economics, in Rotterdam, the Netherlands. He teaches international transport and logistics, maritime economics, and policy and developments in global freight transport BSc and MSc, and he is also involved in postexperience courses. He is Academic Director of the MSC in Customs and Supply Chain Compliance. He participates and coordinates research projects on global supply chains and international logistics and customs supervision in the international movement of goods.

Introduction

Dong-Wook Song

1 Beginning . . . .

Adaptability is all about the survival. This is particularly true in the business world including the maritime sector. Maritime business is volatile in nature, largely due to the fact that the supply side is clumsy: that is, less-responsive to unexpected demands, and relatively more time to take back to the right track towards a so-called equilibrium between demand and supply (Ma, 2020). This very nature of volatility becomes even more capricious and unpredictable by the currently ongoing pandemic, causing the maritime business world to be exposed to the greater uncertainty. Recent years will be definitely remembered a special or unprecedented incident by human history in terms of scale and impact on every aspect of human life. Throughout the history, however, human beings have been all the time seeking out a way to respond to the (un)expected challenges.

The historian E. H. Carr (1961) describes this phenomenon as a never-ending process of challenges and responses. This book could be regarded as a 'collective' response to the unexpected challenges imposed on every aspect of today's maritime business.

D.-W. Song (✉)
World Maritime University, Malmo, Sweden
e-mail: ds@wmu.se

B.-W. Ko, D.-W. Song (eds.), *New Maritime Business*, WMU Studies in Maritime Affairs 10, https://doi.org/10.1007/978-3-030-78957-2_1

2 About the Book

This book is composed of the four Parts under the headings of 'Recent Market Development', 'Decarbonization and Green Finance', 'Digitalization and Maritime Big Data', and 'Technology and New Business Model'. Each Part has two to three chapters having discussed topics related to the heading.

In the Part I, which contains three chapters. The chapter 'Volatility and Uncertainty in Container Shipping Market' of Lam, Li and Pu examines the volatility and uncertainty in the shipping industry and takes a special look into the container shipping market under the background of the currently ongoing COVID-19 pandemic. The Port of Singapore, one of the busiest container ports, has experienced a certain level of impact during the pandemic period. Their analysis focuses mainly on Singapore's container shipping market, such as container throughput, container shipping traffic, and freight rate performance. The performance of container shipping implied by the case port data is analyzed to illustrate the very nature of market volatility and uncertainty. The chapter also conducts a comparison between the container shipping sector and other cargo shipping sectors in Singapore. Findings from the analysis show what had happened during the COVID-19 crisis, and suggest the necessity towards digitalization and automation in the industry.

Johns investigates in the chapter 'Containerships and Black Swans: The Impact of Disruptive Events on the Industry' that the behavior and reactions of container shipping lines to two unforeseen crisis events that had occurred in a time span of a dozen years, the financial and economic crisis following the Lehman collapse in 2008/2009 and the COVID-19 pandemic in 2020. He frames both events as "black swans" and observes similarities and differences that occurred especially in the pricing of transportation services provided by container shipping lines and their impact on the wider container shipping market in general. Based on market data, the chapter shows how the container liner companies have evolved over a little more than a decade and how they changed their behavior, approached the markets differently and reacted in a quicker and more active manner. Furthermore, the chapter attempts to introduce first avenues of explanation that may be the underlying factors for such a behavioral adaptation.

Chapter 'COVID-19: Challenges and Future Responses in International Logistics', produced by Park and Kim, aims at presenting interesting interpretations and key words useful to building a new framework in the field of international logistics and its relevant industrial sectors in response to the uncertainty and challenges mainly caused by the outbreak of COVID-19 pandemic, probably the most disruptive and unprecedented risk imposed on the global supply chain in the recent years. The chapter consists of three parts: literature review focusing on the pandemic as a supply chain risk (SCR) source; examining the impacts of COVID-19 on global economies and international logistics; and, a preliminary exploration of a new approach towards mitigating SCR. By doing so, this chapter argues that a SCR mitigation strategy, specifically focused on impacts of the pandemic as a SC disruptor in global scale, has been relatively underdeveloped. Having made a few key descriptions of the COVID-19 impacts on the global supply chains, of companies' responses in global market, and of experts' observations and analyses, the authors attempt to distill the differences of nature and mechanism of SC disruption between pre- and post-COVID-19, which can be characterized by combination of supply as well as shocks, and rapid move of business operations to the "contactless" world, replacing at least partly efficiency with resilience as priority, while human factors playing a most unique and formidable role especially when encountering the SC disruption. In order to find the bottom line for building a new framework for SC mitigation strategy specific to pandemic-related SC disruption, the authors introduce "key words" being re-emerged and/or discussed in relevant research fields; they include preparedness, resilience and agility, digitalization, and SC reconfiguration. In their concluding part, Park and Kim urge that all of the key words require great attentions from international business world as well as serious consideration from governmental and public bodies in a way of smoothly navigating this ever-changing and -growing uncertain business world.

The Part II has the two chapters. In the chapter 'Container Shipping Decarbonization Pathways', Cariou and Lindstad discuss one of the critical issues in the maritime community—that is, shipping decarbonization. The International Maritime Organization agreed in 2018 on a strategy to support the long-term goal of a reduction in maritime Greenhouse gas (GHG) emissions by at least 50% by 2050. To achieve this ambitious and grand objective, short- and long-term goals were set up. First, to reduce carbon intensity measured in CO_2 per ton-mile by at least 40% by 2030, pursuing efforts toward 70% by 2050, compared to 2008. Second, to reduce the total annual GHG emissions by 2050 by at least 50%, compared to 2008. This chapter assesses whether these two objectives are going to be met by 2030 and 2050 with the case of the container shipping segment, which in total accounts for approximately 30% of GHG shipping emissions by merchant vessels. The authors estimate actual and future carbon intensity using changes over time from a linear extrapolation per ship size, and cautiously conclude that the goals on carbon intensity reduction will probably be met by 2030 and 2050. Moreover, they use forecasts on future trade developments so as to predict future container transport work and to again carefully conclude that the objective of halving emissions by 2050 is within reach for container shipping, only if trades increase less than 1.5% per year. This projection looks interesting to be seen as we are all keen on reducing those harmful emissions.

Chapter 'Green Shipping Finance: Existing Initiatives and the Road Ahead', prepared by Kavussanos and Tsouknidis, discusses the developing regulatory framework on green shipping finance and the role of capital providers in reducing the environmental footprint of the industry. It is well known that transportation services provided by the global shipping industry contribute substantially to the phenomenon of climate change through vessel greenhouse-gas (GHG) emissions. Several initiatives suggested by regulatory bodies and private companies are examined, with the purpose of reducing shipping-related emissions through green financing schemes. The chapter does also make a critical discussion on the progress made so far and the road ahead.

The Part III is consisted of three chapters under the heading of Digitalization and Maritime Big Data. In the chapter 'Maritime Digital Business Models: The Case of D/S Norden, 2015–2020', Iversen examines the digitalization of the maritime sector with a specific focus on business models. It is argued that current research in Maritime Informatics is focused more on technological optimizations and less on commercial aspects in a way to grasp the importance of digitalization in the shipping sector. As per the identified gap, a business model framework is suggested with focus on the level of 'customer-based-value-propositions' and 'land versus sea', respectively. The author moves to the empirical case of the Danish shipping company Norden and its development made from 2015 to 2020 from a perspective of digitalization-driven business. The Norden is a leading commercial operator of dry bulk and product tanker vessels with more than 350 vessels in operation. The conclusion of this case study is that the Norden has regarded digitalization as a tool for decision-taking processes, which, in the long-term, should lead to a position where the company enjoys a competitive advantage, in terms of efficient and effective decision-makings, based on big data and advanced algorithms. The shipping company has on the other hand decided not to use digitalization for the development of new software product and in accordance to presented digitalization matrix focused on indirect value proposition for the customers, rather than direct customer-based initiatives. This focused strategy confirms the hypothesis that digitalization in the dry bulk and tanker segments will often be based on indirect value propositions, while digitalization in container-shipping might have more directly linked to specific customer-based value propositions. This segment-based distinction is closely associated with the business-to-business nature in the dry bulk and tanker markets, and with the mixed business-to-business and business-to-consumer nature of container shipping, in particular, when the container shipping is integrated to the value chains and thus moved closer to the ultimate customers' preferences and services. This line of preliminary findings will surely lead researchers in the field to a number of comprehensive and rigorous empirical tests over the years to come.

Chapter 'Platform Revolution in Container Shipping Markets: An Economics Perspective', written by Ko, investigates a so-called 'platform revolution' in container shipping markets from a perspective from economics. Drawing on a comprehensive economic literature on platform expansions, the author argues that there are (and will be) enormous economic benefits extractable from platform revolution in general and container shipping markets in particular. For the exposition for this argument, the chapter reviews the literature in a way to conceptualize the platform phenomena, having focused on the network economics. Second, it suggests the two interesting and important examples of platforms in the global container shipping markets. One is the blockchain platforms, such as TradeLens and GSBN. The other is the platform-related market price index, such as FBX. Then, this chapter looks into the prospective benefits of, barriers to, and threats from the platform revolutions in container shipping markets. Finally, this chapter concludes by suggesting that the market players should act at the earliest possible time for the adaptation to this rapid digitalization process, including platform revolution, so as not to be expelled from the market but to develop and prosper in the market for the future.

The last chapter in the Part III, chapter 'Liner Services in Asia's Major Container Ports: Implications for Busan', delivered by Sahoo, proposes a simple port intra-dependency ratio to measure the degree (or percentage) of which a particular port depends on reference port(s) in the adjacent region, having in mind co-existence of co-operation and competition between ports. As the international trade of commodities is crucial for the economic development of any nation or region, ports are regarded as the gateway to the development of any country. Since Asia is the world's manufacturing hub, this chapter investigates the trading pattern of container line services for three major ports in Asia: that is, Busan, Singapore and Shanghai. As container lines provide services in closed loops, multiple ports are called at during their services. In addition, ports compete for common markets as there are overlaps in ports' hinterlands and forelands. The study utilizes the automatic identification system (AIS) and liner services data from AXS Marine's Alphaliner for the necessary analysis. Findings from the examination suggest that the Port of Shanghai generates almost 26% of its volume without depending on Busan and Singapore, while Singapore gets almost 40% of liner capacity without depending on Busan and Shanghai. However, Busan generates only 11% of liner capacity on its own: that is, without relying upon Singapore and/or Shanghai. These figures indicate that the Port of Busan is considered heavily dependent on Singapore and Shanghai for its container services, thus, creating its increasingly risky cargo-handling profiles. The Busan Port is expected (or required) to diversify its liner services and routes portfolios and to expand its self-dependency in a way to become a major transshipment hub in the Far East. This chapter concludes by identifying some challenges and opportunities faced with by the Port of Busan as a case study using a region-based maritime big data set. Big Data Analytics in the maritime industry could act as a decision support system for stakeholders, in this case, port managers and policymakers, to constantly realize the market's gaps and needs as a manner to materialize them, thus increasing the port market share and retaining and/or sustaining container throughputs.

The final Part IV has two chapters. Kim and Schröder-Hinrichs in their chapter 'Research Developments and Debates Regarding Maritime Autonomous Surface Ship: Status, Challenges and Perspectives' discuss recent developments in automation and telecommunication technologies with the prospect of autonomous ships, associated research domains, and regulatory matters related to Maritime Autonomous Surface Ships (MASS) development and operations. Propelled by several MASS projects and trials in recent years, the said field is evolving rapidly from a distant dream to a near reality by virtue of substantial interests and R&Ds. New operational concepts based on remote-control and autonomous vessel functions are projected to increase efficiency and reliability, and to reduce operating expenditures and GHG emissions, while having great potentials to contribute towards a range of safety and economic benefits to the maritime industry in general and shipowners in particular. This chapter offers a review of recently conducted studies on autonomous ship operations with the aim to contribute to ongoing discussions with regard to the following two aspects. Firstly, it attempts to synthesize the current research discussions and debates on MASS. Secondly, it explores possible barriers and hurdles on the way to putting MASS in full operation. The chapter does also suggest future research directions.

Chapter 'New Business Models for Shipping: Innovation in the Netherlands', of Veenstra, provides an overview of innovations and research in three main areas: zero-emission shipping, digital and autonomous shipping, and safe shipping. The shipping industry is treated in the chapter as a playground for innovation in technology and business models. The author brings our attention to the Dutch maritime innovation agenda. In all identified areas in the Netherlands, stakeholder or parties in concern are conducting research and developing new technologies and solutions as a form of business and operational innovation. Two cases are discussed in a detailed manner, where start-up companies have emerged out of this R&D effort, leading to the stage that new business models in the shipping industry are being created.

3 .... Ending

All those chapters are concerned with the single challenge facing the maritime business world—that is, uncertainty. At the same time, each chapter deals with a specific area of the maritime business community in an effort to better understand the complicate markets, to seek for a solution of economic or financial sustainability under the pressure of climate changes, to discuss the technology as an option for the future, and finally to show how to utilise the big data set for better informed decision- and policy-makings that used to be unfeasible in terms of scale and capacity. It is hoped that all those endeavours manifested in the book are considered as the first and small step towards practically transforming the industry in line with Schumpeter (1943), and academically changing a paradigm of thinking and scientific discovery in line with Kuhn (2012).

References

Carr, E. (1961). *What is history?* Palgrave Macmillan.
Kuhn, T. (2012). *The structure of scientific revolutions* (4th ed.). University of Chicago Press.
Ma, S. (2020). *Economics of maritime business.* Routledge.
Schumpeter, J. (1943). *Capitalism, socialism and democracy.* Routledge.

Suggested Reading

Baden-Fuller, C., & Stopford, J. (1994). *Rejuvenating the mature business: The competitive challenges* (2nd ed.). Harvard Business School Press.
Levinson, M. (2020). *Outside the box: How globalization changed from moving stuff to spreading ideas.* Princeton University Press.

Mazzucato, M. (2014). *The entrepreneurial state: Debunking public vs. private sector myths*. Anthem Press.
McCraw, T. (2007). *Prophet of innovation: Joseph Schumpeter and creative destruction*. Harvard University Press.
Nonaka, I., & Takeuchi, H. (1995). *The knowledge-creating company*. Oxford University Press.
Senge, P. (2006). *The fifth discipline: The art and practice of the learning organization*. Random House.
Sundararajan, A. (2016). *The sharing economy: The end of employment and the rise of crowd-based capitalism*. MIT Press.

Part I
Recent Market Development

Volatility and Uncertainty in Container Shipping Market

Jasmine Siu Lee Lam, Qingyao Li, and Shuyi Pu

Abstract This chapter studies the volatility and uncertainty in the shipping industry and takes a special look into the container shipping market under the background of the COVID-19 pandemic. The Port of Singapore, which is one of the busiest container ports, has experienced a certain level of impact during the pandemic. The analysis focuses mainly on Singapore's container shipping market regarding container throughput, container shipping traffic, and freight rate performance. The performance of container shipping implied by the port data in Singapore is analysed to illustrate the volatility and uncertainty in the market. The study also conducts a comparison between the container shipping sector and other cargo shipping sectors in Singapore. With the results shown what had happened during the COVID-19 crisis, the implications reveal the volatile nature of the shipping markets and suggest the necessity of developing digitalisation and automation in the maritime industry.

1 Introduction

According to the World Economic Forum (WEF), the global economy faces rising uncertainty and has seen signals of downward revision (WEF, 2019). The trend could be complicated with multiple factors, such as international political situations (e.g. the escalating trade tensions between US and China, and Brexit), economic stress in countries (e.g. Argentina and Turkey), and disruptions to industries caused by new regulations (e.g. the auto industry in German is facing new emission regulations). The situation could get even more uncertain in 2020 and in the near future. Look at what has happened in 2020: armed conflicts existed between some countries which brought regional disorders; inconclusive factors such as the 2020

J. S. L. Lam (✉) · Q. Li · S. Pu
School of Civil and Environmental Engineering, Nanyang Technological University, Singapore, Singapore
e-mail: SLLam@ntu.edu.sg

B.-W. Ko, D.-W. Song (eds.), *New Maritime Business*, WMU Studies in Maritime Affairs 10, https://doi.org/10.1007/978-3-030-78957-2_2

US election added more uncertainties to the global economy. The World Uncertainty Index has experienced a sudden rise in 2020 (IMF, 2020), which indicates worries of growing uncertainties faced by most countries in the world. Shipping, as a derived demand heavily dependent on the global economy, is therefore vulnerable to such uncertainties and no doubt being affected.

Apart from the uncertainties from the external environment as mentioned above, shipping markets are also exposed to the uncertainties from internal factors in the industry. Examples are the stricter regulations for the industry in terms of safety, environment and quality such as decarbonisation regulations (Ship Technology, 2019) and business strategies of participators in the industry such as alliances' reshuffles in container shipping lines (McKinsey, 2016). These uncertainties could affect the market sentiment that leads to adjustments of behaviours in the market (Notteboom & Lam, 2014).

Among many internal and external factors, the COVID-19 pandemic has shown significant impacts on the shipping industry (IMO, 2020), and signifies the feature of volatility and uncertainty of the shipping market. As an infectious disease, the COVID-19 has become a pandemic affecting numerous countries globally (WHO, 2020) and has been even called the COVID-19 crisis. Mapping the timeline of COVID-19, the first reported cluster of cases of pneumonia was on 31 December 2019, following by the first outbreak news on 5 January 2020 (WHO, 2020). Since then, it has spread quickly and has been assessed as a pandemic in March 2020. The outbreak and global spread of the COVID-19 pandemic have placed a more negative impact than people previously anticipated. Singapore is also hit by this pandemic, with more than 50,000 confirmed infection cases and has adopted circuit breaker measures as well as phased approaches of re-opening economic activities to fight against it. According to the Ministry of Trade & Industry (MTI) in Singapore, the COVID-19 pandemic has brought about severe disruptions to global economic activities and led to both demand-side and supply-side shocks to the Singapore economy and various industries (MTI, 2020). The Port of Singapore, which is one of the busiest container ports in the world, has also experienced a certain level of impact during the pandemic. Particularly for the shipping industry, the impact of COVID-19 penetrates different areas, such as maritime trade, crew change, maritime safety, and security. It is important to analyse these impacts and learn from them for future references. Therefore, this study aims at studying the volatility and uncertainty in container shipping related to COVID-19. The data analysis focuses mainly on Singapore's container shipping market. The performance of container shipping implied by the port data in Singapore is analysed to illustrate the volatility and uncertainty in the market. The study also conducts a comparison between the container shipping sector and other cargo shipping sectors in Singapore.

The remaining of the study is structured as follows. Section 2 reviews related literature. Section 3 presents the analysis of volatility and uncertainty in the container shipping market from various perspectives. Section 4 discusses implications and recommendations to stakeholders. At last, the conclusion and contributions are summarised in Sect. 5.

2 Literature Review

The shipping industry is volatile and uncertain in nature (Notteboom & Lam, 2014). Many studies have been conducted to analyse the volatility and uncertainty of shipping markets from different aspects. Firstly, some researchers focused on understanding the patterns of volatility in the shipping industry. Hsiao et al. (2014) investigated the freight volatility transmission between dry bulk and container shipping markets associated with the 2008 financial crisis by using the GARCH-BEKK model. They found that the dry bulk sector has short-term spillover effects on the container sector during the crisis period, vice versa after the crisis. Pouliasis et al. (2018) studied the volatility dynamics and risk transmission of the stock price in different shipping sectors including container, dry bulk, tanker, and gas. The results suggested a strong correlation in large losses and supported asymmetric transmission among those sectors. Ranjan et al. (2018) analysed the volatility of bunker markets with E-GARCH and bivariate analysis. Their analysis showed that both IFO 380 and IFO 180 have short- and long-term shocks persistence and observed a causal relationship between IFO380 and IFO 180 bunker prices.

Beyond understanding the industry's volatility patterns, some studies aimed to develop or compare models that factor in different uncertainties in the industry for risk assessment and decision making. Nielsen et al. (2014) developed a forecasting model embedded with the uncertainty of the container shipping market and analysed the influence of time horizon and observation fit on forecast results. Bai and Lam (2019) modelled shipping destination choice behaviours with the analysis of market uncertainties like product price arbitrage and bunker price. The study focused on the very large gas carrier market which showed the model's applicability in transport planning. Rodrigues et al. (2019) compared thirteen modelling techniques with uncertainty in travel time for a maritime inventory routing problem given constant production and consumption rates. The results proved that by considering uncertainty in the model setup, substantial gains could be realised. They also provided guidance on which approach is better with different considerations.

Another trend of literature in the research area is about analysing the influencing factors of the volatility and uncertainty in shipping markets or the volatility affected by certain exogenous events. Homan (2007) studied the impact of the Maritime Transportation Security Act on the volatility of marine firms using the market model. The result indicated that the Act could reduce the financial risks of marine operators and have positive effects on maritime companies in terms of raising capital. Chi and Cheng (2016) analysed the impact of exchange rate volatility and real income on Australia's maritime export flows to its major trading partners like China. They found that the volatility of the exchange rate was an important factor that had a long-run effect on maritime export volume. Thanopoulou and Strandenes (2017) proposed a framework to classify and analyse risks and uncertainties in shipping and estimated the potential ranges of future shipping until 2040. They suggested that it is necessary to focus on the nature of consequences for the shipping industry to increase resilience. Specifically, regarding the impact of COVID-19 on shipping market fluctuation, there is one paper from Michail and Melas (2020) which examined the volatility in dry bulk, clean and dirty tanker sectors caused by COVID-19. Their analysis showed a direct impact of COVID-19 on dry bulk and dirty tanker markets, with second-round effects via a decline in oil prices and sometimes third-round effects via the stock market.

The above analysis reveals that there are scarce studies on the volatility of the shipping industry in relation to random shocks such as the outbreak of pandemic. Therefore, this study aims to analyse the volatility and uncertainty in container shipping related to COVID-19 with a focus on Singapore. The analysis will be conducted from various aspects of the shipping market including port throughput, vessel arrival, and freight rate.

3 Volatility and Uncertainty in Container Shipping in Relation to COVID-19

The following analysis focuses on Singapore's container shipping market. As one of the world's busiest ports, the Port of Singapore ranks as the second in terms of annual container throughput in 2020 (Lloyd's List, 2020a). Singapore is an island nation's port, which means the demand of the nation itself is not a major pillar of the container throughput. Instead, majority of the throughput comes from transshipment, marking the Port of Singapore a transshipment hub in the Southeast Asia region and connecting the traffic around the world. In 2018, the shipping volume in Southeast Asia rises at a fast pace in the container trade (Lloyd's List, 2018). In 2019, there was a tendency of a shift of production centres from China to other Asian countries affected by the trade war between China and the US, and the Port of Singapore still acted as a hub for transshipment (Lloyd's List, 2019). Container shipping is a very important sector for the business and operation in the Port of Singapore. In the following analysis, the data derived from the Maritime and Port Authority of Singapore (MPA) of Singapore is used to examine the volatility and uncertainty in the container shipping market.

3.1 Insights from Container Throughput in Singapore

The container throughput fluctuated in the first half of 2020 with the emergence of COVID-19. Figure 1 shows the pattern of monthly throughput in twenty-foot equivalent units (TEUs) in 2019 and 2020. There was a slight reduction of 0.55% in January 2020 compared to the throughput in December 2019 and later on a big drop in the throughput volume in February was explicit. In 2020, the monthly throughput dropped from 3,182,600 TEUs in January to 2,898,700 TEUs in February, which accounts for a 8.92% decrease. Following in March, an increase

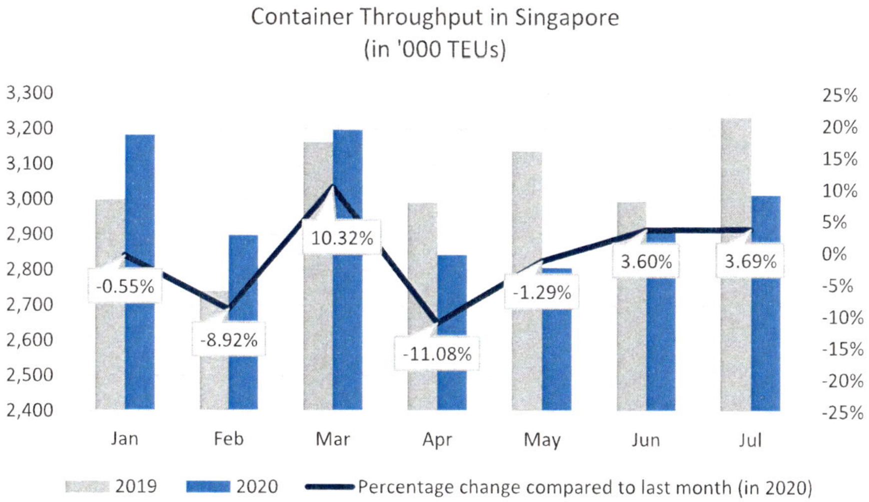

Fig. 1 Container throughput in Singapore from January to July in 2019 and 2020. Source: Drawn by authors, based on data from MPA

in throughput occurred, with a 10.32% rise of volume compared to February, which brings the throughput to almost the same level as in January. However, the positive change did not hold given that the volume reduced to 2,843,500 TEUs in April with an 11.08% decrease from March, marking the biggest monthly drop in 2020. In May, the throughput hit the trough with 2,806,700 TEUs. The continuous volume drop from March to May signalled a depressed performance of the market. The throughput realised a stable recovery since June 2020, although it is still below the pre-pandemic level. As mentioned earlier, Singapore is a transshipment hub, thus throughput in Singapore reflects the demand of the region. Therefore, such recovery in throughput since June 2020 could in some degree indicate the demand revival in the region.

As for containerised cargo throughput by weight, the trend also shows a fluctuating and volatile pattern, as shown in Fig. 2. From January to June 2020, the total weight of containerised cargo throughput in tons in Singapore experienced a reduction in February, rebounded in March, and continued to decline in April and May and then gradually recovered. The pattern of containerised cargo throughput change in tons is very similar to the pattern of container throughput in TEU, therefore, later when comparing the year-to-year change, the study focuses on the container throughput for analysis.

Figure 3 depicts the year-to-year trend of container throughput by percentage change compared to the same month last year in the past 5 years. The red line is the trend of change in 2020. It is found that the throughput in 2020 is generally lower than that in 2019. Although the year-to-year monthly change from 2019 to 2020 shows a positive value, the percentage change tends to be gradually moderate in the first three months, from 6.12% in January, to 5.77% in February, and 1.07% in March. The downward pattern of the trend happened in the Port of Singapore, and

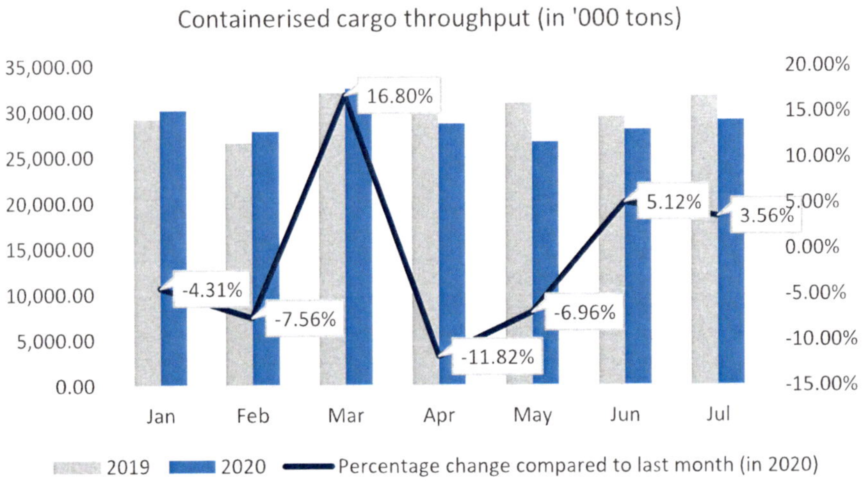

Fig. 2 Containerised cargo throughput in Singapore from January to July in 2019 and 2020. Source: Drawn by authors, based on data from MPA

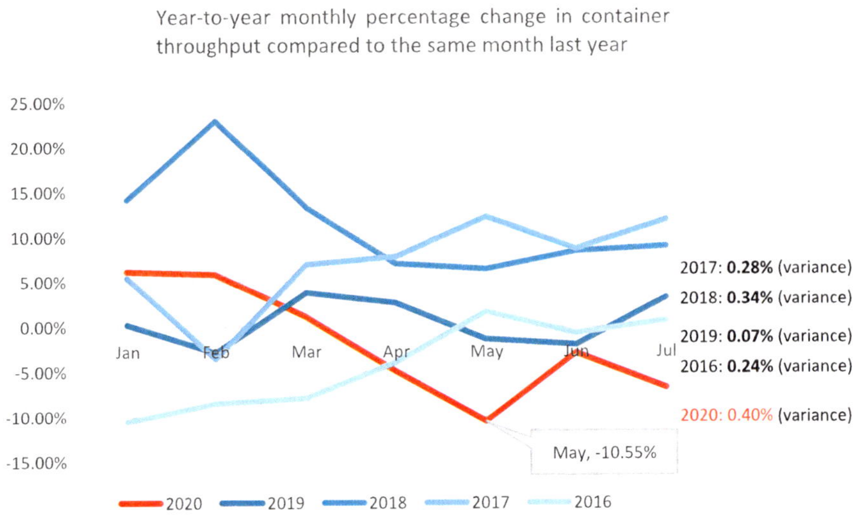

Fig. 3 Year-to-year monthly percentage change. Source: Drawn by authors, based on data from data.gov.sg

such a decreasing tendency in the first quarter of 2020 was also seen in the rest of the world. For example, according to the International Transport Forum (ITF), in February 2020 the global container trade volumes declined by 8.6% compared to the same month last year (ITF, 2020). From April to July 2020, the monthly throughput in Singapore all remains less than the same monthly throughput comparing to last year, revealing a decreasing trend in year-to-year monthly percentage change. The lowest throughput volume was in May, which was −10.55% less than the same month in 2019.

Overall, 2020 is the most volatile year in the recent 5 years. By calculating the variance of the year-to-year monthly percentage change in container throughput from January to July, it is noteworthy that the variance in 2020 reached 0.40%, while the variances of the past 4 years are smaller, ranging from 0.07% to 0.34%. This implies a larger degree of volatility and uncertainty in 2020 when the COVID-19 pandemic occurs.

3.2 *Insights from Container Vessel Arrival in Singapore*

Another perspective to investigate the container shipping performance is to examine the number of arrivals of container vessels. According to what is available from MPA's data, container vessel arrivals by number and by gross tonnage are used to generate insights on the status of container shipping in Singapore. Generally, the number of cargo ships calling at the port of Singapore declined greatly in 2020 than the previous years, especially in May when the number dropped to 3059, the lowest since at least 1993 (The Straits Times, 2020). The container vessel calls also experienced a downward trend during those few months.

For container vessel arrivals in Singapore by number in 2020, the pattern is similar to the container throughput. Figure 4 shows the number of container vessel arrivals in Singapore from January to July in 2019 and 2020. The number of vessels

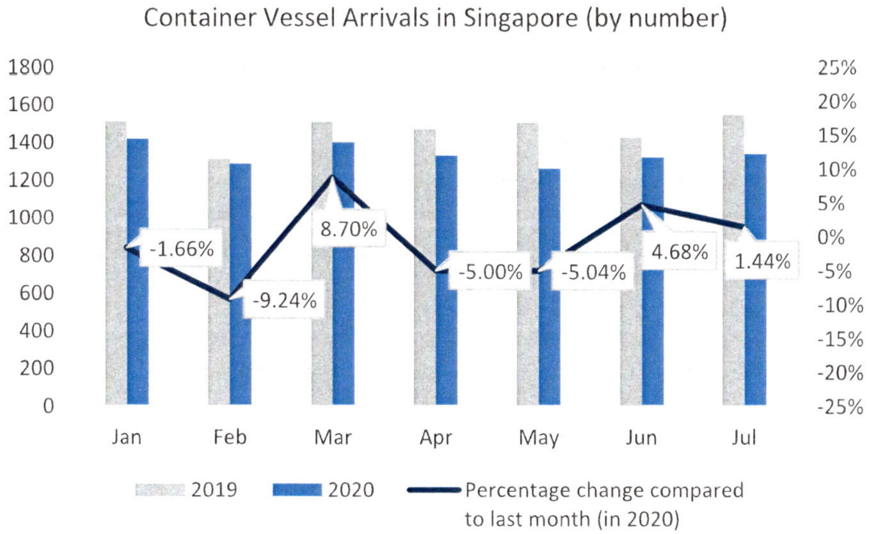

Fig. 4 Container vessel arrivals in Singapore from January to July in 2019 and 2020 (by number). Source: Drawn by authors, based on data from MPA

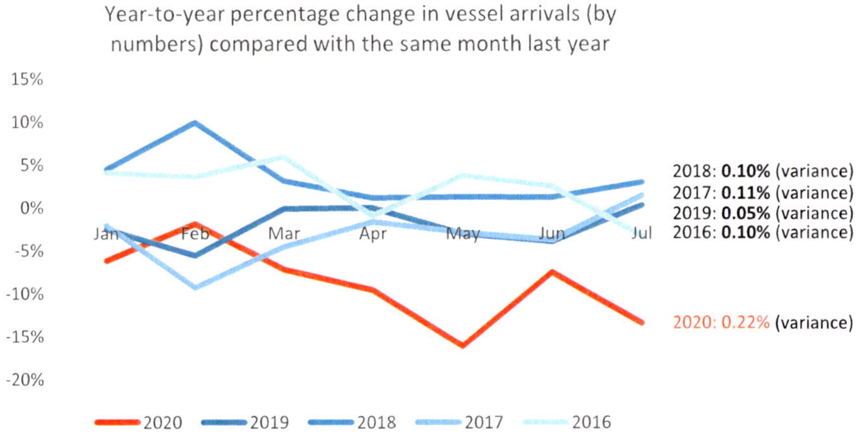

Fig. 5 Year-to-year monthly percentage change in the number of vessel arrivals. Source: Drawn by authors, based on data from data.gov.sg

fluctuated in the first few months and then started to remain in a slightly growing status since June. Compared to January, the number of vessel arrivals fell to 1287 in February, dropped by about 9.24%. Then there was a transient increase in March that the number of arrivals reached almost the same level as the number in January. However, in April and May, the number of ships kept relatively low in arrivals, especially in May that the number reached the lowest being 1262. This indicates a significant reduction in the supply of container vessel capacity.

When comparing monthly year-to-year data, Fig. 5 shows the year-to-year monthly percentage change in the number of vessel arrivals in the past 5 years. It is found that the number of vessels in 2020 was lower than that in 2019 every month from January to July, with the largest reduction of more than 16% being in May as shown in Fig. 5. The average year-to-year monthly reduction reached 8.91% compared to the previous year. The trend in Fig. 5 also indicates that 2020 is the most volatile year in the recent 5 years with the biggest variance in the percentage change of vessel arrivals by number compared with the same month last year. The variance of the year-to-year monthly change is 0.22% in 2020, while the previous 4 years' variance is around 0.1%.

For container vessel arrivals in Singapore by gross tonnage, as shown in Fig. 6, the lowest gross tonnage was in February, which was 12.7% lower than the previous month. If compared to the same month in 2019, the gross tonnage in February 2020 yet showed a small increase of 2.83%. The lowest year-to-year percentage change happened in May that the gross tonnage has experienced a change of −9.30%. June and July remained a negative trend compared to the same month last year, and the recovery seems slow and insignificant. The container vessel arrivals by gross tonnage, together with the number of container vessel arrivals in Singapore, have shown that although vessel arrivals gently increased since June, the total capacity provided by the vessels still remained at a relatively low level. One of the reasons for

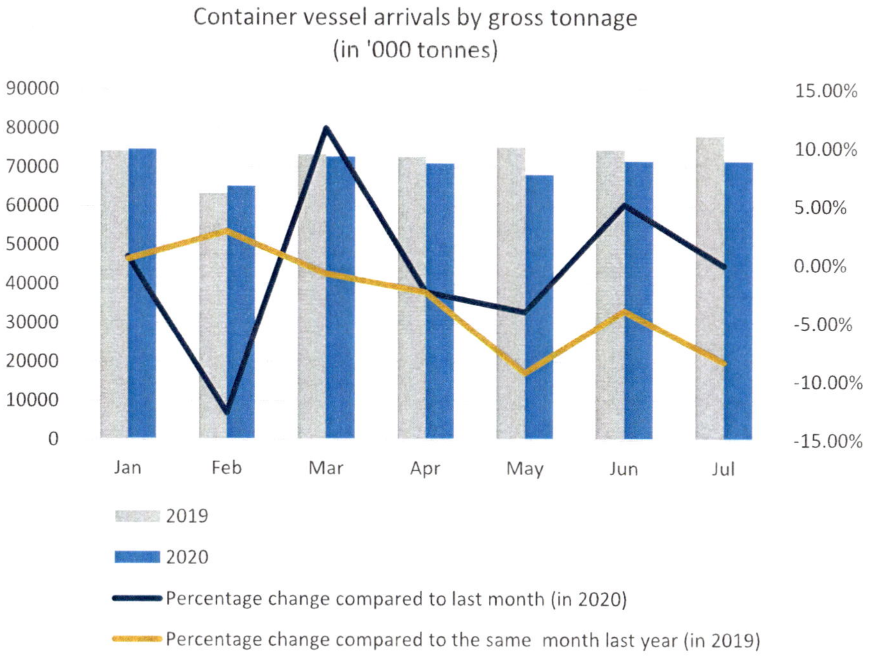

Fig. 6 Container vessel arrivals by growth tonnage. Source: Drawn by authors, based on data from MPA

the reduced on-sail capacity is the cancellation of sails and lay-up of the ships, which will be explained in the next part.

3.3 Insights from Container Freight Rate

Referring to Shanghai Containerized Freight Index (shown in Fig. 7) and World Container Index (shown in Fig. 8), the freight rate generally decreased in the first four months since the outbreak of COVID-19 in China in early January 2020. Since April, the freight rate has stably improved and even exceeded the pre-pandemic level in May. Later in July and August, the freight rate has experienced robust growth and the Platts key container freight rates hit record highs (Hickin & Griffiths, 2020). The main reason for the volatility in freight rate is the continuous change of demand and the adjustment of supply.

As we have analysed earlier, in the port of Singapore, the container shipping volume in general has decreased in January to May 2020. The rest of the world also suffered from declining in cargo demand. For the far east region, while the COVID-19 outbreak first happened in China in January 2020, there was a −17.5% change in container trade volume in the far east region in February 2020 (ITF, 2020). The

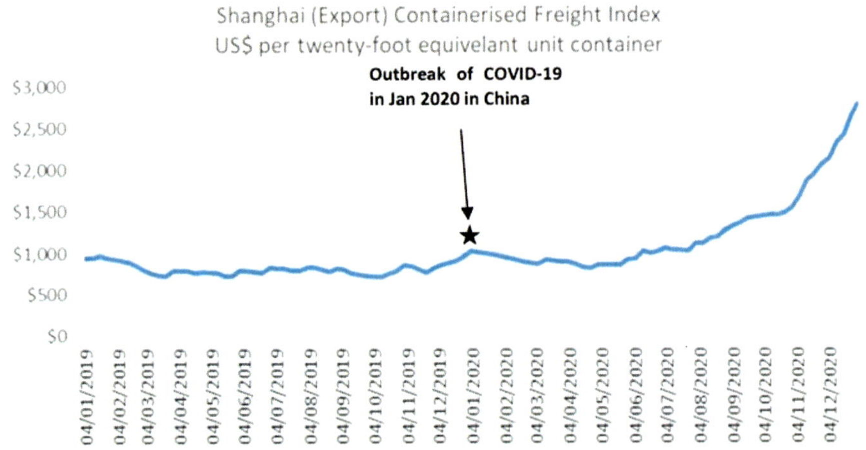

Fig. 7 Shanghai containerized freight rate from 2019 to 2020. Source: Drawn by authors, based on data from Bloomberg

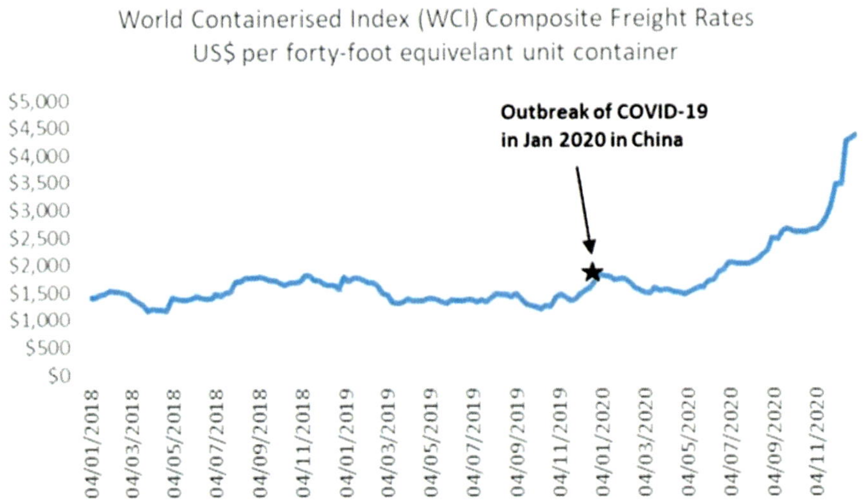

Fig. 8 World Container Index 2018 to 2020. Source: Drawn by authors, based on data from Bloomberg

drastic drop in container trade volume was due to the supply shock in manufacturing countries such as China and demand shock in other countries that adopted lockdown measures to prevent the spread of the virus. Starting from late January, many countries locked down the cities with serious infections and thus the trade demand generally declined around the world due to slower economic activities. Shipping plays an important role in global trade, thus the impact of decline in trade demand is directly reflected in the shipping market. With the decreased shipping demand and overwhelmed supply of vessel capacity, the freight rate decreased from January to late April 2020.

Responding to the demand-crushing effects of the pandemic, shipping companies adopted aggressive vessel capacity reduction strategies by blank sailings, which refer to cancellation of sailings by skipping certain ports that were originally included in the routes or cancelling the entire journey (The Straits Time, 2020). The shipping alliances increased blank sailings to match the demand shock. In 2020, till August, the three major alliances, 2M, THE and Ocean[1] have voided more than 400 sailings (Hickin & Griffiths, 2020). According to the International Transport Forum, there were 85 blank sailings in the Asia-North America West Coast trade lane and 49 on the Asia-North Europe trade lane, and in total, the blank sailings reached 188 in February and March 2020 (ITF, 2020). The pattern of blank sailings reveals that there was an increase in the number of blank sailings in the first quarter of 2020, especially for the Asia-North America West Coast lane where the sudden increase in blank sailings was spotted (ITF, 2020). From January to April 2020, although the shipping alliances have adopted the blank sailings strategy to reduce the supply of capacity, the demand and supply of shipping were still not matched. As the demand was still below the supply these months, the freight rate was decreasing from January to late April 2020. With time going on, the capacity adjustment continued, and the blank sailings gradually reduced the capacity significantly. Till the middle of June, the total pandemic-induced blanked capacity was nearing four million TEUs (The Business Time, 2020). Through August 2020, there were 126 blank sailings in the Trans-Pacific route and 94 in the Asia-Europe route from the three shipping alliances. Through September, more than 70 sailings were cancelled. The total shipping capacity was reduced by more than 50% in some large volume trade routes (JOC, 2020a). When the capacity provided by shipping lines continued to reduce, the decrease in supply gradually caught up with the decrease in demand. This resulted in a soar in container freight rate. Moreover, with the gradual recovery and the end of lockdowns in many countries, the shipping demand is rising. However, the supply of shipping capacity is yet back to normal because the carriers are still flexing capacity to match the demand affected by COVID-19 (JOC, 2020b). Therefore, the rates remain at a high level but the future trend is still enclosed by uncertainties. Another aspect of freight rate volatility and uncertainty reflects in the regional differences. Although the growth in freight rate is generally seen in most major shipping routes especially the far-east to US route since late April 2020, for some routes such as the spot rates to Europe did not perform well and dropped significantly in April. From May, the Asia-Europe freight rate moderately increased, but still fell behind the Asia-America rate. This indicates some degree of heterogeneity of freight rate around the world and the uncertainty across different regions (BIMCO, 2020).

[1] 2M Alliance: Maersk and MSC; Ocean Alliance: COSCO, OOCL, CMA CGM, and Evergreen; THE Alliance: Hapag-Lloyd, ONE, & Yang Ming.

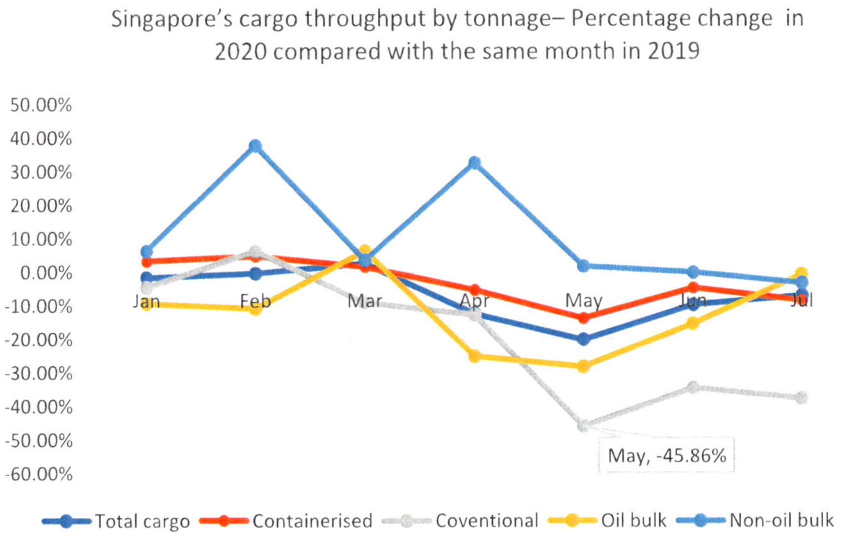

Fig. 9 Cargo throughput percentage change in 2020 compared with the same month in 2019 by different cargo sectors. Source: Drawn by authors, based on data from MPA

3.4 Comparison with Other Shipping Markets in Singapore

This section compares the performance of the container shipping market and other cargo shipping markets including conventional cargo, dry bulk and oil bulk. Generally speaking, the total cargo volume in Singapore in 2020 experienced an average of −6.97% monthly change compared to the same month last year, and the lowest throughput was in May, with 44.6 million tons, which is 20.10% lower than May 2019. Figure 9 shows the cargo throughput by the tonnage percentage change in 2020 compared with the same month in 2019 for different sectors of cargo types. As shown in Fig. 9, almost all shipping sectors present negative trends compared to the same month last year, except for the non-oil bulk market. The amount of throughput of non-oil bulk remained steady and almost unchanged from January to April 2020. The year-to-year monthly change of non-oil bulk also held to be positive compared to the same month last year from January to May. From June to July, the throughput slightly decreased compared to the same month last year and the year-to-year monthly change became negative. Different from the performance of non-oil bulk, the conventional cargo shipping market, however, was affected the most and performed the worst among all the cargo types. The volume of conventional cargo drops by a significant 45.86% in May, a notable 34.49% in June and 37.59% in July compared to the same month in 2019. As for the degree of volatility for different cargo types, the trend in Fig. 9 implies that the conventional cargoes, non-oil bulk and oil bulk are the top three volatile markets in terms of cargo throughput, with the variance of the change compared to the same month in 2019 being 3.27%, 2.38%

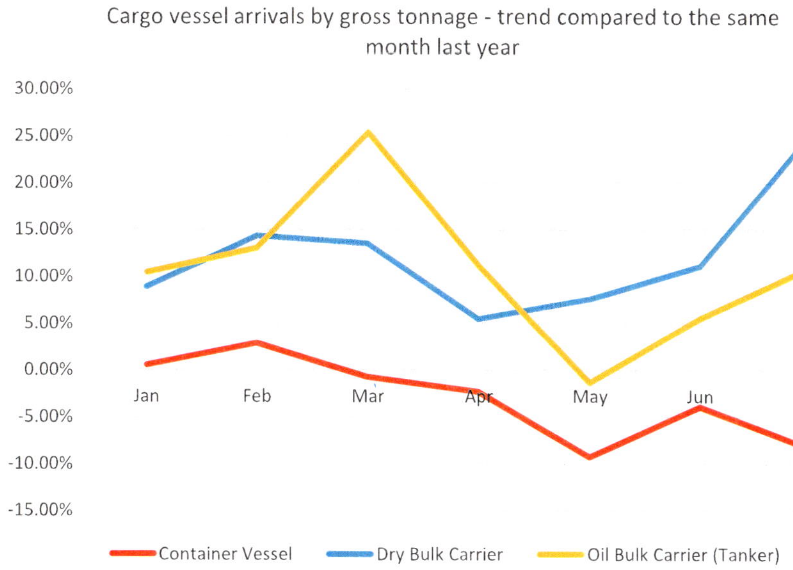

Fig. 10 Cargo vessel arrivals percentage change in 2020 compared to the same month last year by different carriers. Source: Drawn by authors, based on data from MPA

and 1.31%. Our analysis finds that the container shipping market is relatively less volatile than other markets, with the variance being 0.39%.

When looking at shipping traffic, we are interested in the three major cargo shipping sectors, namely container vessels, dry bulk carriers, and tankers which are comparable. Dry bulk carriers correspond to the previously mentioned non-oil bulk cargoes while tankers correspond to oil bulk cargoes. Figure 10 shows the trend of percentage change compared to the same month last year for cargo vessel arrivals by gross tonnage. For the average monthly percentage change in vessel arrivals by gross tonnage in 2020 compared with the same month in 2019 from January to July, it is found that the gross tonnage of container vessel arrivals decreased by 3.1%, while bulk carrier increased by 10.7%, and tanker increased by 12.3%. Similarly, when considering the average monthly percentage change of vessel arrivals by numbers in 2020 compared with the same month in 2019, the data shows container vessels had an average of 8.91% decrease, while bulk carriers had an 8.84% increase, and the tankers had a 10.84% increase. It is noticed that the positive percentage change in bulk carriers' arrivals compared to the same month last year corresponds to the generally positive performance in non-oil bulk throughput. However, while the arrivals of tankers also show a positive increase compared to the same month in 2019, the oil cargo throughput did not reflect an increase. This behaves differently from the container shipping market, where we see both the cargo throughput and the ship arrivals display a general decreased trend compared to the same month last year. The synchronisation reflects the aggressive capacity adjustment to match the demand and supply of containerised cargoes. Besides, the container shipping market is less volatile than bulk shipping and tanker, where the variances were 0.17%, 0.37%, 0.56% respectively in terms of gross tonnage and 0.19%, 0.57%, 0.53% in terms of the number of arrivals. This aligns with the previous finding that non-container shipping markets are more volatile in terms of cargo throughput.

When comparing the freight rate, it seems that the container market, dry bulk market and tanker market are all volatile in 2020. Dry bulk appears to be more uncertain than the container market, despite a similar decreasing trend in Jan-May 2020. Since May 2020, the container freight rate has been steadily increasing but the Baltic Dry Index still fluctuates. For the tanker market, referring to the tanker market index by Clarksons Platou, the rate was more unstable in 2020 and there was a sudden increase in April (Clarksons Platou, 2020). The boom in April for the tanker market was mainly due to a surge in demand for tankers as floating oil storage increased. According to Lloyd's list, floating storage volume reached the highest in record in 2020 since 2009 (Lloyd's List, 2020b). This was caused by low oil price due to weak global crude oil consumption. Meanwhile, the contangoes on the oil market also flourished. In the near future, when the demand for oil recovers, the tanker rates would still face great uncertainties in view of many external factors.

4 Implications and Recommendations

In 2020, most sectors in shipping have encountered difficulties and even crisis under the shadow of COVID-19. The vulnerability features of shipping were exposed to stakeholders in the maritime industry. As one of the major sectors in the shipping market, container shipping also faced with volatility and uncertainty. The chapter studies the performance and reactions of the container shipping market during the COVID-19 pandemic period, and the implications derived from the analysis shall give the industry certain suggestions and recommendations to face future volatility and uncertainties.

Back to the nature of shipping as a derived demand, the different status of demand and supply could affect the situation of the shipping market and bring volatility and uncertainty to the industry. The sudden reduction in demand in international trade at the early stage of the pandemic incurred the drop in container shipping demand. Such status was reflected in the declined container throughput and the decrease in container freight rate. At this point, the responsive supply adjustment by blank sailings was effective to counter volatility and uncertainty in the container shipping market by matching the supply to the decreased demand. The container freight rate recovered fast and even exceeded the pre-COVID-19 level due to collaborated capacity adjustment in container shipping alliances. Consequently, although the shipping volume was going down, the major container shipping lines such as Hapag Lloyd, AP Moller-Maersk, HMM, and Ocean Network Express among others have reported a positive net profit in the second quarter of 2020 (Abdi, 2020). From this point of view, the freight rate reflects the adjustment on the relationship between shipping demand and supply. Meanwhile, the large-scale cancellations of sailings could also affect the shipping schedules and their reliability, thus Drewry has suggested shippers and forwarders check the situation carefully before booking and reserve enough lead time for themselves when anticipating operational delays (Drewry, 2020a).

Besides the insight from the freight rate, the year-to-year monthly change in container throughput and container vessel arrivals in the port of Singapore has the smallest variance among major cargo shipping sectors, and the trend in the container throughput has shown a signal of recovery. People have seen the resilience of container shipping during the crisis. Besides, the resilience of the Port of Singapore against uncertainty and risk helps container shipping in the region to survive and recover. As shipping demand is volatile, the resilience to confront the volatility is very important. When the demand dropped during the early pandemic period, many ports faced difficult financial and operational conditions as the blank sailings caused the vessels to skip the ports. For example, the Asia-Europe trade lane was affected significantly due to the demand shock and the ports in the Mediterranean that originally host many ultra-large container ships were hit heavily (Ovcina, 2020a). This intensified the competition among the ports in the same region. Therefore, taking the port of Singapore as an example, to remain the advantage as a hub port and stand out from the ports in the same region requires better competitiveness and performance, such as prompt and appropriate reactions when facing shorter notice on changes of shipping schedules. At the same time, when the demand recovers, the port also needs to react quickly to the change. In the post-pandemic period, facing possible cargo surge, the port shall be flexible and capable to handle the operations. In addition, the special safety measures during the COVID-19 also help to control uncertainties and remain resilient. Those measures were applied to ensure the smoothness and safety of crew change, thereby facilitating shipping companies in this crucial aspect. For example, the "safe-corridor" procedures were imposed by MPA to safeguard the crew change process in Singapore. This includes the need to apply Stay-Home Notice and the direct transfers of crew (Seatrade Maritime News, 2020).

With a relatively high level of automation and attached attention to digitalisation, during the pandemic period, the Port of Singapore still remains in operation. For automation in the Port of Singapore, there are many projects and research in developing port technologies related to automation, one of which is a collaborated program by PSA and MPA in 2012, where SGD 20 million was funded into it and the content included test-bedding projects on Automated Guided Vehicle (AGV). The purpose to apply such automated equipment and facilities is to reduce the reliance on human labor and increase productivity in the port (Lam, 2016; Tay, 2012). Such attempts of automation lead to greater resilience when facing the uncertainties and unexpected results brought up by crises such as the COVID-19 pandemic. During the COVID-19, when the crew onboard or staff in the terminal fall ill, the normal operation of vessels and handling of cargoes will be at a standstill if the reliance on physical manpower is too high and the integration of the handling process is inadequate. For example, vessels have experienced delays in India during the COVID-19 crisis because the pilot failed in going aboard the vessel. Also, with the health concerns when the crew have been travelled to high-risk countries previously, many ports would like to accelerate the process of stepping into remote pilotage to reduce the risks. The remote control carried out by skilled employees without a physical presence at a specific location and the application of fully autonomous equipment are in favor to be seen in the future (JOC, 2020c). Full automation is still a long way to go. During the pandemic, Yara Birkeland, the first autonomous and electric container vessel in the world has been paused in May 2020 due to the COVID-19 pandemic (Ovcina, 2020b). Although nowadays autonomous ships which are fully controlled by systems are still at an early stage in the business application, increasing the level of automation will indeed have plenty of spaces to benefit the industry.

At the same time, considering crew safety, operation, and management efficiency, and uplifting the level of automation, the lessons learned by the shipping industry during COVID-19 indicate the need to accelerate digitalisation in maritime trade and maritime supply chains. Amid unexpected shocks, digitalisation would aid the industry to counter similar uncertainties and risks in the future. Organisations are more resilient to disruptions caused by pandemic events if they are able to provide digital services. Operations can be transferred from office to home while maintaining good service quality with a high level of digitalisation. Digital solutions are considered more critical by maritime stakeholders in post-COVID-19 operations and this requires joint efforts from the private and the public sectors. COVID-19 has in some sense drives the development of digitalisation in the container shipping sector as the traditional in-person business becomes difficult and even impossible during the pandemic period. Furthermore, digitalisation brings about potential cost reduction and value-added benefits such as easier communication channels for conducting businesses, and those features propel the industry going more digital.

Specifically, blockchain is a promising technology to address the digitalisation need in the industry while maintaining good data integrity, security, transparency, and traceability (Lam et al., 2020; Pu & Lam, 2020). In the post-pandemic environment, blockchain can help to rebuild the disrupted networks with transparency, trust and security provided to the trading partners, and to integrate the physical and digital flow in the supply chain which is also connected to the financial transaction (WEF, 2020). One of its applications is the blockchain bill of lading. For instance, partnering with the Infocomm Media Development Authority, MPA is embarking on the development of blockchain-based electronic bills of lading in Singapore, known as TradeTrust initiative. This initiative aims for the secure exchange of digital trade documents in the maritime sector and reduction of trade processing and documentation costing up to 20% of the physical cost of shipping a single container (Riviera, 2020). As a whole, with the results shown what had happened during the COVID-19 pandemic, the implications reveal the volatile nature of the shipping market and suggest the necessity of developing digitalisation and automation in the maritime industry.

Meanwhile, focusing on innovations by tapping on emerging technologies could also bring lights to the container shipping industry when facing volatility and uncertainties. A study conducted by Lam and Zhang (2019) revealed that the top three innovative solutions for enhancing customer value of liner companies are the use of eco ship and eco container technology, big data solution as well as automation and digitalisation of systems. For example, as unusual disruptions could take place during COVID-19, on a weekly basis, the Drewry Containership Waiting Time Tracker utilising the Automated Identification System (AIS) data calculates and updates the average waiting times for container ships outside a port before berthing at 44 major ports (Drewry, 2020b). At the port side, new technologies and innovations which incorporate automation, data analytics and robotics etc. are addressed for catering container vessels and meeting the complex requirements of shipping lines. To build the next generation mega port in Tuas, PSA in Singapore has embraced new technologies for container terminals such as amphibious drones, automated quay cranes, exoskeletons for port staff, and robotic arms for related container activities (The Business Times, 2018). The innovative attempts also consider utilising the features of multimodal transportation, which is seen as one of the growth areas that has emerged from the global pandemic. Singapore is considering developing and improving the abilities in sea-air transshipment, in order to deliver cargo in a timely manner, especially for chilled products (Channel News Asia, 2020). However, investing in and developing such emerging technologies themselves also involve potential risks and uncertainties, thus the stakeholders shall be more cautious when tapping into the new technologies and innovations.

In addition, when thinking about how to be resilient in the volatile and uncertain environment, the sustainable business models and strategies are the cornerstones to support the survival and development of the container shipping industry. Scholars suggest a hybrid business model taking care of both cost minimisation and service differentiation and innovation (Lam & Wong, 2018), which enables flexible and versatile capabilities of liner companies when facing different market environments. Therefore, the hybrid business model is crucial in an industry surrounded by volatility and uncertainty. Taking the example of what happened to the liner companies during the COVID-19, the cost for a shipping company is particularly important when encountering crises. Healthy financial status matters to the survival of the business. However, in the post-pandemic period when the demand goes up, the differentiation of the services will support the company to stand out from the competitors. For example, although the cargo demand generally declined in the COVID-19 pandemic, there is still a considerable need for transporting some special products such as medical and material products. Such demand will last for a period of time even in the post-pandemic period with the potential increase in the need of vaccines and other medical products. Therefore, providing supply chain solutions in handling special products with good and reliable services will help the companies to be more competitive in the industry.

5 Conclusion

In this chapter, we present an analysis of the volatility and uncertainty of the container shipping market, especially focusing on the performance of the Port of Singapore. The analysis of the volatility and uncertainty is conducted from several perspectives, including the performance of container throughput, vessel arrivals and freight rate. Singapore's container shipping market has experienced the most volatile environment in the recent 5 years during the COVID-19 period. The market is always under the impact of the changes in demand and supply. In Singapore's container shipping sector, the container throughput in 2020 has generally dropped compared to the previous year, meanwhile, the vessel arrivals have also decreased in number and in gross tonnage. There were drastic ups and downs in the volume change, indicating a certain degree of volatility in the industry. But when compared to other cargo shipping sectors such dry bulk and tankers, container shipping volume remained less volatile than those sectors. As for the freight rate, the change in the rate reflects the relationship between shipping demand and supply. The drop in cargo demand at the beginning stage of the pandemic pulled the freight rate down, but later the aggressive shipping capacity adjustment such as blank sailings by shipping alliances pushed the freight rate up that even exceeded the pre-pandemic level. In a word, although volatility and uncertainty exist at almost every aspect of the container shipping market, so far the container shipping market has displayed considerable resilience against the volatile and uncertain environment.

Facing the unpredictable and inevitable changes in the market, in the short run, the container shipping companies mitigated the volatility and uncertainty by taking measures to balance the demand and supply, for example, dealing with the unexpected decrease in freight rate during COVID-19 caused by shipping demand reduction with fast response in blank sailings. In the long run, some of the effective ways to survive such crises and increase resilience to uncertainties include leveraging automation and digitalisation such as blockchain, adopting innovations and new technologies, as well as considering a hybrid business model to increase a company's flexibility and resilience. A high level of automation and digitalisation aids to promote the remote control on the operations with less manpower necessarily in presence, which will increase the safety level in the industry. This is particularly crucial during crises like pandemic. In addition, operation efficiency and cost efficiency will also be uplifted with the application of automation and digitalisation.

The chapter provides several contributions to the field of studying the volatility and uncertainty of the container shipping market. Firstly, it is an early attempt to study the impact of COVID-19 on the container shipping market focusing on the situation of the Port of Singapore. The analysis serves as a reference to better understand the volatility and uncertainty of the market. Secondly, it provides implications for container shipping companies and governments on how to accelerate a sustainable recovery and build resilience capabilities. Although many ports around the world, including the Port of Singapore, have been through a tough time, and the shipping lines also faced with great demand uncertainties, the container shipping market has generally survived and revealed its resilience against the crisis. One of the emphases learned from the experience during COVID-19 is the benefit and necessity to develop digitailisation and automation in the container shipping industry. Thirdly, besides addressing the difficulties that occurred during the COVID-19 crisis, the chapter also points out the possible future path and opportunities for the industry to evolve and develop. In the future, other studies may dig more into the topic of volatility and uncertainty in container shipping and other shipping sectors, for the sake of providing more insights in dealing with unexpected events.

References

Abdi, B. (2020). Capacity management to help container shipping liners double profits in 2020. S&P Global Platts. Retrieved October 10, 2020, from https://www.spglobal.com/platts/en/market-insights/latest-news/shipping/090320-capacity-management-to-help-container-shipping-liners-double-profits-in-2020

Bai, X., & Lam, J. S. L. (2019). A destination choice model for very large gas carriers (VLGC) loading from the US Gulf. *Energy, 174*, 1267–1275. https://doi.org/10.1016/j.energy.2019.02.148

Baltic and International Maritime Council. (2020). Container shipping: Massive blanking of sailings has supported freight rates as demand collapses. Retrieved September 18, 2020, from https://www.bimco.org/news/market_analysis/2020/20200526_container_shipping

Channel News Asia. (2020). Sea-air freight model an opportunity to uplift Singapore as a transport hub: Experts. Retrieved September 18, 2020, from https://www.channelnewsasia.com/news/business/singapore-transport-hub-sea-air-freight-13075634

Chi, J., & Cheng, S. K. (2016). Do exchange rate volatility and income affect Australia's maritime export flows to Asia? *Transport Policy, 47*, 13–21. https://doi.org/10.1016/j.tranpol.2015.12.003

Clarksons Platou. (2020). Tanker Market Index. Retrieved September 30, 2020, from https://www.nat.bm/wp-content/uploads/2020/07/Weeklyfreightrates200703.pdf

Data.gov.sg. (2020). Maritime and Port Authority of Singapore datasets. Retrieved October 30, 2020, from https://data.gov.sg/dataset?organization=maritime-and-port-authority-of-singapore

Drewry. (2020a). Sailing schedules fall victim to COVID-19. Retrieved October 09, 2020, from https://www.drewry.co.uk/logistics-executive-briefing/logistics-executive-briefing-articles/sailing-schedules-victim-of-COVID-19

Drewry. (2020b). Drewry begins tracking weekly containership cancelled sailings and waiting times. Retrieved October 02, 2020, from https://www.drewry.co.uk/news/drewry-begins-tracking-weekly-containership-cancelled-sailings-and-waiting-times

Hickin, P., & Griffiths, G. (2020). Shipping alliances help container industry boost freight rates in coronavirus pandemic. S&P Global Platts. Retrieved October 10, 2020, from https://blogs.platts.com/2020/09/16/shipping-alliances-container-industry-freight-rates/

Homan, A. C. (2007). The impact of MTSA on financial risk and volatility of marine firms. *Maritime Policy & Management, 34*(1), 69–79. https://doi.org/10.1080/03088830601103459

Hsiao, Y. J., Chou, H. C., & Wu, C. C. (2014). Return lead–lag and volatility transmission in shipping freight markets. *Maritime Policy & Management, 41*(7), 697–714. https://doi.org/10.1080/03088839.2013.865849

International Maritime Organization. (2020). Coronavirus disease (COVID-19) Pandemic. Retrieved October 12, 2020, from https://www.imo.org/en/MediaCentre/HotTopics/Pages/Coronavirus.aspx

International Monetary Fund. (2020). World Economic Outlook Update, June 2020. Retrieved October 02, from https://www.imf.org/en/Publications/WEO/Issues/2020/06/24/WEOUpdateJune2020

International Transport Forum. (2020). COVID-19 Transport Brief: Global Container Shipping and the Coronavirus Crisis. Retrieved September 30, 2020, from https://www.itf-oecd.org/sites/default/files/global-container-shipping-COVID-19.pdf

Journal of Commerce. (2020a). COVID-19: Alliances Outline Extensive Blank Sailings for Q3. Retrieved September 30, 2020, from https://www.joc.com/maritime-news/alliances-outline-extensive-blank-sailings-q3_20200603.html

Journal of Commerce. (2020b). Asia-Europe container rates to remain elevated through Q4: Service providers. Retrieved September 30, 2020, from https://www.joc.com/maritime-news/asia-europe-container-rates-remain-elevated-through-q4-service-providers_20200911.html

Journal of Commerce. (2020c). COVID-19 to accelerate digitalization, automation of container shipping. Retrieved September 30, 2020, from https://www.joc.com/international-logistics/COVID-19-accelerate-digitalization-automation-container-shipping_20200326.html

Lam, J. S. L. (2016). Strategy of a transhipment hub: The case of port of Singapore. In *Dynamic shipping and port development in the globalized economy* (pp. 12–38). Palgrave Macmillan. https://doi.org/10.1057/9781137514295_2

Lam, J. S. L., Goh, P. G., & Pu, S. (2020). *Impact of disruptive technologies on maritime trade and maritime industry*. Nanyang Technological University and National University of Singapore.

Lam, J. S. L., & Wong, H. N. (2018). Analysing business models of liner shipping companies. *International Journal of Shipping and Transport Logistics, 10*(2), 237–256. https://doi.org/10.1504/IJSTL.2018.090078

Lam, J. S. L., & Zhang, X. (2019). Innovative solutions for enhancing customer value in liner shipping. *Transport Policy, 82*, 88–95. https://doi.org/10.1016/j.tranpol.2018.09.001

Lloyd's List. (2018). One Hundred Ports 2018. Retrieved October 04, 2020, from https://lloydslist.maritimeintelligence.informa.com/one-hundred-container-ports-2018#:~:text=One%20Hundred%20Ports%202018&text=The%202018%20edition%20of%20Lloyd's,growth%20on%20the%20previous%20year

Lloyd's List. (2019). One Hundred Ports 2019. Retrieved October 04, 2020, from https://lloydslist.maritimeintelligence.informa.com/one-hundred-container-ports-2019

Lloyd's List. (2020a). One Hundred Ports 2020. Retrieved October 04, 2020, from https://lloydslist.maritimeintelligence.informa.com/one-hundred-container-ports-2020

Lloyd's List. (2020b). Floating Storage at Record Levels as Tanks Fill up on Land. Retrieved September 16, 2020, from http://lloydslist.maritimeintelligence.informa.com.remotexs.ntu.edu.sg/LL1132091/Floating-storage-at-record-levels-as-tanks-fill-up-on-land

Maritime and Port Authority of Singapore. (2020). Port Statistics. Retrieved September 18, 2020, from https://www.mpa.gov.sg/web/portal/home/maritime-singapore/port-statistics

Michail, N. A., & Melas, K. D. (2020). Shipping markets in turmoil: An analysis of the Covid-19 outbreak and its implications. *Transportation Research Interdisciplinary Perspectives, 7*, 100178. https://doi.org/10.1016/j.trip.2020.100178

Ministry of Trade & Industry in Singapore. (2020). Impact of the COVID-19 pandemic on the Singapore economy. Retrieved September 16, 2020, from https://www.mti.gov.sg/-/media/MTI/Resources/Economic-Survey-of-Singapore/2020/Economic-Survey-of-Singapore-First-Quarter-2020/FA_1Q20.pdf

Nielsen, P., Jiang, L., Rytter, N. G. M., & Chen, G. (2014). An investigation of forecast horizon and observation fit's influence on an econometric rate forecast model in the liner shipping industry. *Maritime Policy & Management, 41*(7), 667–682. https://doi.org/10.1080/03088839.2014.960499

Notteboom, T., & Lam, J. S. L. (2014). Dealing with uncertainty and volatility in shipping and ports. *Maritime Policy & Management, 41*(7), 611–614. https://doi.org/10.1080/03088839.2014.965297

Offshore Energy. (2020a). European ports hit the hardest by blanking of sailings. Retrieved October 09, 2020, from https://www.offshore-energy.biz/european-ports-hit-the-hardest-by-blanking-of-sailings/

Offshore Energy. (2020b). Development of the world's 1st zero emissions, autonomous feeder halted due to COVID-19. Retrieved October 09, 2020, from https://www.offshore-energy.biz/development-of-the-worlds-1st-zero-emissions-autonomous-feeder-halted-due-to-COVID-19/

Pouliasis, P. K., Papapostolou, N. C., Kyriakou, I., & Visvikis, I. D. (2018). Shipping equity risk behavior and portfolio management. *Transportation Research Part A: Policy and Practice, 116*, 178–200. https://doi.org/10.1016/j.tra.2018.06.016

Pu, S., & Lam, J. S. L. (2020). Blockchain adoptions in the maritime industry: A conceptual framework. *Maritime Policy & Management.* https://doi.org/10.1080/03088839.2020.1825855

Ranjan, I., Raju, T. B., Jayaraj, R., & Dutta, K. (2018). Understanding volatility and leverage effects in bunker markets. *International Journal of Global Energy Issues, 41*(1–4), 158–175. https://doi.org/10.1504/IJGEI.2018.092310

Riviera. (2020). Blockchain-based bill of lading initiative launched by Singapore authorities. Retrieved September 25, 2020, from https://www.rivieramm.com/opinion/blockchain-based-bill-of-lading-initiative-launched-by-singapore-authorities-21521

Rodrigues, F., Agra, A., Christiansen, M., Hvattum, L. M., & Requejo, C. (2019). Comparing techniques for modelling uncertainty in a maritime inventory routing problem. *European Journal of Operational Research, 277*(3), 831–845. https://doi.org/10.1016/j.ejor.2019.03.015

Saxon, S. (2016). The alliance shuffle and consolidation: Implications for shippers. McKinsey. #. Retrieved October 02, 2020, from https://www.mckinsey.com/industries/travel-logistics-and-transport-infrastructure/our-insights/the-alliance-shuffle-and-consolidation-implications-for-shippers

Seatrade Maritime News. (2020). Singapore beefs up safety measures for crew change. Retrieved October 09, 2020, from https://www.seatrade-maritime.com/ship-operations/singapore-beefs-safety-measures-crew-change

Ship Technology. (2019). Sink or swim: How decarbonisation will reshape shipping forever. Retrieved October 02, 2020, from https://www.ship-technology.com/features/how-will-decarbonisation-impact-shipping/

Tay, E. (2012). PSA Invests in Automation Technologies. Retrieved November 10, 2020, from http://www.internationalpsa.com/psanews/pdf/nr120716.pdf

Thanopoulou, H., & Strandenes, S. P. (2017). A theoretical framework for analysing long-term uncertainty in shipping. *Case Studies on Transport Policy, 5*(2), 325–331. https://doi.org/10.1016/j.cstp.2017.03.008

The Business Times. (2018). PSA Singapore unveils advanced port technologies in new exhibition. Retrieved September 30, 2020, from https://www.businesstimes.com.sg/transport/psa-singapore-unveils-advanced-port-technologies-in-new-exhibition

The Business Times. (2020). Cargo ship calls at Singapore hub plummet to 27-year low. Retrieved October 02, 2020, from https://www.businesstimes.com.sg/transport/cargo-ship-calls-at-singapore-hub-plummet-to-27-year-low

The Straits Times. (2020). Cargo ship arrivals in Singapore plunge to 27-year low in May. Retrieved October 02, 2020, from https://www.straitstimes.com/business/economy/cargo-ship-arrivals-in-spore-plunge-to-27-year-low-in-may

World Economic Forum. (2019). The Global Economy Is Being Hit by Higher Uncertainty. Retrieved September 18, 2020, from https://www.weforum.org/agenda/2019/05/the-global-economy-hit-by-higher-uncertainty/

World Economic Forum. (2020). This is how blockchain can be used in supply chains to shape a post-COVID-19 economic recovery. Retrieved September 18, 2020, from https://www.weforum.org/agenda/2020/06/this-is-how-blockchain-can-be-used-in-supply-chains-to-shape-a-post-COVID-19-economic-recovery/

World Health Organization. (2020). Archived: WHO Timeline - COVID-19. Retrieved September 20, 2020, from https://www.who.int/news-room/detail/27-04-2020-who-timeline%2D%2D-COVID-19

Further Reading

Lam, J. S. L., & Bai, X. (2016). A quality function deployment approach to improve maritime supply chain resilience. *Transportation Research Part E, 92*, 16–27. https://doi.org/10.1016/j.tre.2016.01.012

Zhang, X., & Lam, J. S. L. (2019). A fuzzy Delphi-AHP-TOPSIS framework to identify barriers in big data analytics adoption: Case of maritime organizations. *Maritime Policy & Management, 46*(7), 781–801. https://doi.org/10.1080/03088839.2019.1628318

Lloyd's List. (2018). One Hundred Ports 2018. Retrieved October 04, 2020, from https://lloydslist.maritimeintelligence.informa.com/one-hundred-container-ports-2018#:~:text=One%20Hundred%20Ports%202018&text=The%202018%20edition%20of%20Lloyd's,growth%20on%20the%20previous%20year

Containerships and Black Swans: The Impact of Disruptive Events on the Industry

Max Johns

Abstract This paper is investigating the behavior and reactions of container shipping lines to two unforeseen crisis events that had occurred in a time span of a dozen years, the financial and economic crisis following the Lehman collapse in 2008/09 and the COVID-19 pandemic in 2020. We frame both events as "black swans" and observe similarities and differences that occurred especially in the pricing of transportation services provided by container shipping lines and their impact on the wider container shipping market. Based on market data we will show how the container liner companies have evolved over a little more than a decade and how they changed their behavior, approached the markets differently and reacted faster and more actively. Furthermore, we try to introduce first avenues of explanation that may be the underlying factors for such behavioral adaptation.

1 Introduction

The notion of the "black swan" has been used since the financial and economic crisis in 2008/09 as a topos to explain how economic players react to unforeseen events and a certain magnitude. Following the influential publication of Nassim Nicholas Taleb (Taleb, 2007) by the same title, economists have learnt to acknowledge extreme outlier events with large impact on business and the economy. In this article we are interested to see if we can derive from data analysis an insight into the container shipping industry and see if they have learnt and adapted over a little more than a decade from one major outlier event (financial crisis 2008/09) to another one (COVID 19, 2020). We use market data to observe reaction patterns, price and volume behavior and respective repercussions for the containership transportation markets.

M. Johns (✉)
Hamburg School of Business Administration, Hamburg, Germany
e-mail: max.johns@hsba.de

B.-W. Ko, D.-W. Song (eds.), *New Maritime Business*, WMU Studies in Maritime Affairs 10, https://doi.org/10.1007/978-3-030-78957-2_3

2 Black Swans Revisited

Nassim Nicholos Taleb has defined "Black Swans" as outstandingly rare events with an extreme impact and retrospective predictability (Taleb, 2007, S. XVIII). With extremely rare events he is describing those that cannot be reasonably predicted. He mentions also those events that would have been extremely likely but did not occur. The strong impact is being brought into the definition as other events would not or barely be noticed. In economic circumstances it would be important to require some observability. The third part of the retrospective predictability is an observation Taleb made around a number of historical black-swan-events. He describes that people have a propensity to explain the formerly unforeseen events back into a reasonable train of thought. Unforeseeable events are not only uncomfortable by their—often negative—impact but also psychologically by the fact that they create uncertainty. Thus, to reduce such uncomfortable uncertainty people tend to create a retrospective predictability. These ex-post explanations are mostly simplistic and even populist. As these outlier events with large impacts have been increasing since the beginning of the industrial revolution, a strange bias evolved: people get better trained in explained events ex-post and are at the same unable to predict them better.

One of the reasons that black-swan-events have significant impacts on businesses and economics developments is the application of risk models. Risk models are necessarily based on past experiences and are extrapolating these into the future. If a certain event has never happened before or has not been included into the group of observations, the models cannot not predict their occurrence. Part of the problem may be also the cognitive bias of the "survivorship bias" (Shermer, 2014; Smith, 2014). Both lead to optimistic overestimations. Rare negative outlier events are not taken into account and create practical protection to discover the possibility of large deviations.

For the chapter we were firstly confronted with the definition of the COVID-19 pandemic as a black-swan-event. All three defining elements had to be fulfilled in order to qualify the events of 2020 and make them comparable to the aftermath of the financial crisis in 2008/09:

- Rarity
- Impact
- Retrospective predictability.

There can be no doubt that the COVID-10 pandemic had an outsized impact on economies around the world. It probably will have had the largest impact outside both World Wars during the last several decades.

There can equally be found numerous accounts of attempts to make the events ex-post more of even highly predictable. Part of the explanations are conspiracy theories that the virus has been intentionally created and set free by one or the other superpowers.

The as easily ticked off is the first qualifier, rarity. There have been impactful virus outbreaks before, namely influenza (Spanish flu 1914, Asian flu 1957/58,

Hong Kong flu 1968/69, Ebola, SARS and MERS. The World Health Organization had described the potential of public health emergencies due to further outbreaks of viral and other diseases as early as 1995 (Council, 2020). For the sake of a good punch line, reinsurer Swiss Re ran whole page advertising campaigns in October and November 2020 in the Economist and other publications reading "Not a Black Swan", with a large picture of a black swan. The brief text explains that COVID-19 was one in a long row of pandemics and can for that reason not qualify as an outlier. Such caustic observation seems to backed up by quotes from Bill Gates who said in a speech at the Massachusetts Medical Society in 2018: "There is a significant probability of a large and lethal, modern-day pandemic occurring in our lifetimes". He voiced similar concerns in 2015.

Most importantly though, Nicolas Taleb investigated the pandemic himself and concluded together with investment manager Mark Spitznagel that the Corona events were not necessarily a black swan (Taleb & Spitznagel, 2020, S. 29). They argue forcefully against aid packages for some industries and decry "corporate socialism" where commercial risk is taken over by governments securing especially the highest salaries and boni. The aim of their intervention is to avoid a confusion between workers who cannot do without government support and companies that the public believes it cannot do without (like airlines). Similarly, he had argued for an article that not corporations but individuals should be bailed out (Avishai, 2020).

If we leave aside arguments of commercial self-interest from Swiss Re and other companies using the potent image of the black swan, the fulfillment of the three components remains important.

Interestingly, Bill Gates' remarks had neither in 2015 nor 2018 any effect. Indeed, he became (ex-post) rather a modern-day Cassandra whose advice was so far outlying that it could neither be digested nor understood by those concerned. As Bill Gates was perfectly capable of reaching out to potent media that may have spread his message more widely, his obvious caution hints at persisting doubts about his own message.

As for the refusal by Taleb himself to categorize COVID-19 as a black-swan-event, he is avoiding the obvious. Neither a dangerous virus nor even a Corona virus have been unpredictable. Both have been known and studied for several decades. However, what qualifies perfectly as a radical outlier event is the economic and social standstill that ensued. A practical global lockdown, several weeks of confinement for entire populations and the largest drop in GDP in history have not been foreseen. Nor predicted. Nor in any way described beforehand. The Black Swan has been the effect, not the cause. Despite numerous studies on exponential growth, exponential break-downs of entire economies have been entirely overlooked. The world was completely unprepared. Thus, from the perspective of the container shipping industry, the effects of COVID-19 in 2020 have clearly been a "black swan" event.

From this observation we believe it is worthwhile to investigate how containerships and black swans have coexisted in 2008/09 and 2020.

3 Volatile Transport Volumes

Container shipping is based on a derived demand. Shipping does not create demand out of itself, it can only transport what the primary industries are producing and shipping. Firstly, we should look at some relative and absolute data (Fig. 1).

The data show that quantitatively there are two very similar events observable in container shipping: From 1998 onwards, there is a clear and almost linear upward trend in the amount of TEU shipped, tonne miles of TEU transported and tonnes of cargo transported in containers. 2009 and 2020 are tow outlier years with negative growth. For 2008 a receding growth is observable due to the negative last quarter. In absolute terms, the drops are similar in both critical years: TEU −12.5m in 2008 and −9.97m in 2020); tonne-miles of TEU −621bn and −504bn respectively; TEU miles −61bn and −54bn and finally −137m tonnes and −93m tonnes transported.

These data show that the world had to similar extents less to transport in absolute terms, a demand shock which had to be absorbed by fleet of rather different sizes.

Setting this into context of GDP developments, compared to world seaborne trade for all types of ships and cargo gives an even more volatile picture (Fig. 2).

The graphs illustrate a significantly different development between both observed black-wan-events in 2009 and 2020. While world seaborne trade turned negative as a consequence of the global financial crisis which morphed into an economic crisis, change in GDP had dropped but remained in positive territory. The COVID-pandemic afflicted GDP much stronger and pushed it into negative growth. Different to what might have been expected, world seaborne trade dropped to a similar degree as in 2009 (−4% and −4.4% respectively).

To understand this phenomenon better, it is important to note that following the financial crisis production industries had been heavily affected. The weight of services in the drop in GDP in 2020 has been much larger and consequently had a lesser effect on seaborne trade which can only transport physical good. The drop in

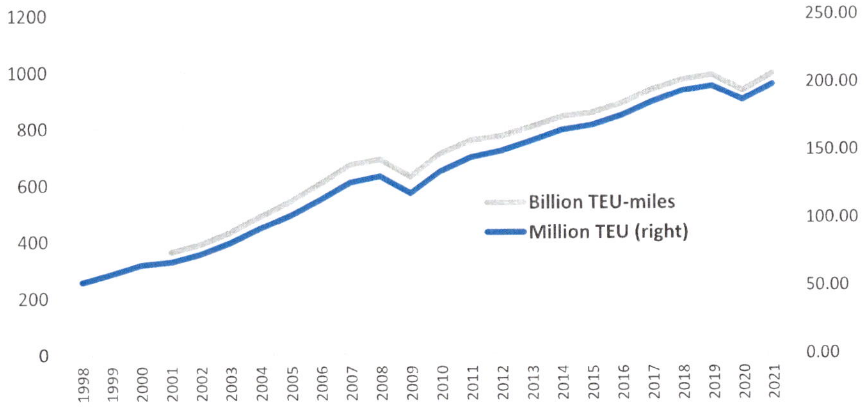

Fig. 1 TEU transported on vessels, measured in million TEU and TEU-miles per year. Source: Clarksons

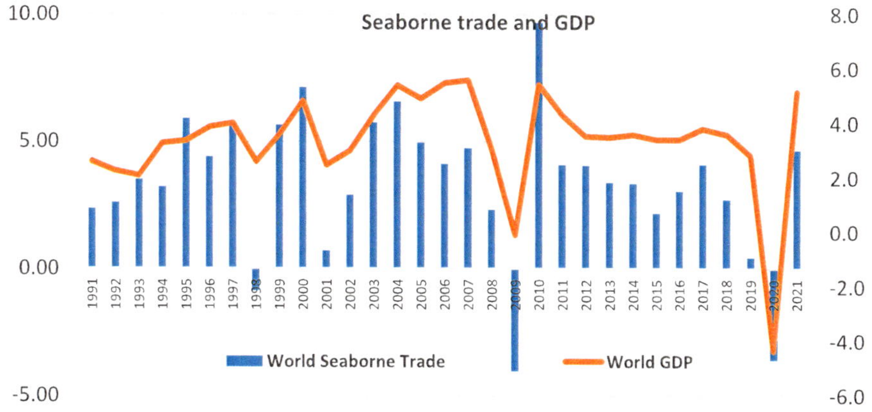

Fig. 2 Relation between seaborne trade and global GDP from 1991 until 2021 (est.). Source: Clarksons

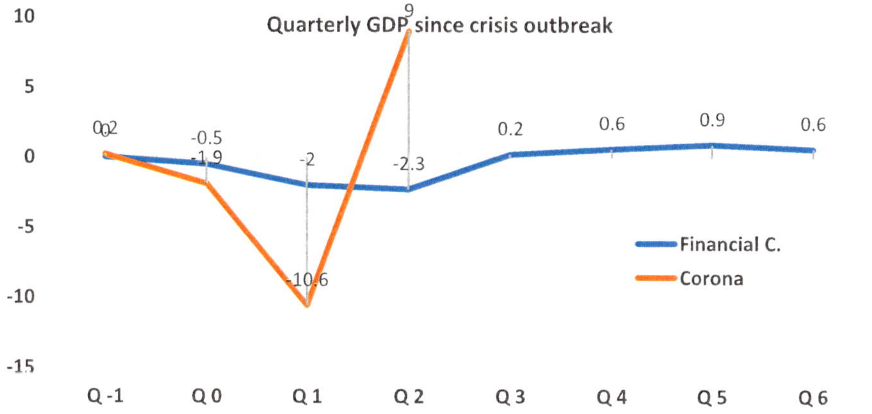

Fig. 3 Development of GDP since the start of both "black swans" in 2008 and 2020. Source. OECD

industrial production was around 30% less around the world compared to 2009. Also, contrary to 2009, it appeared that consumption shifted to a sizable degree from services (holiday travel, eating out, entertainment) to purchases of physical goods (PCs, laptops, home improvement, home office equipment et al.). The latter could lift the amount of cargo to be transported faster than expected (Fig. 3).

Further evidence to a stronger volatility triggered by the two events become visible when plotting OECD growth rates (compared to quarterly data a year earlier) with the outbreak of the crisis (i.e. the default of Lehman Bros. in the fall of 2008 and the publicly known outbreak of the pandemic in the first quarter of 2020). Lock-downs and in particular the temporary early meltdown of economic activities in

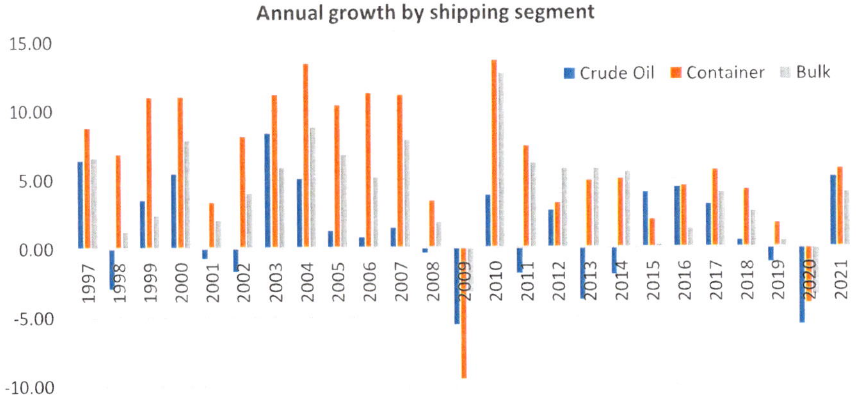

Fig. 4 Comparison of the three major cargo and ship types and their annual growth rates from 1997 until 2021 (est.). Source: Clarksons

China have caused OECD data to drop fast and strongly but rising quickly thereafter. This was somewhat smoother but more protracted in 2008/09.

Shedding now the spotlight onto the shipping industry, we can see that over the last quarter century the different ship and cargo types have fared relatively independently. Taking the three main ship types which make up more than 75% of the maritime cargo capacity, we see a certain similarity between the movements of dry cargoes (container and dry bulk), while wet cargoes meandered in their growth rates often into negative figures (Fig. 4).

As the visual analysis already indicates, the correlation between oil transport and both dry markets is of a positive nature but between 0.52 (container) and 0.62 (dry bulk). Meanwhile, the correlation between both dry cargo types is almost perfect above 0.84. With our focus on both black-swan-events, we can again observe important differentiations. The financial and following economic crisis hit all three sectors, but container-shipping was hit almost twice as hard as both other sectors. In 2020 we see that the energy sector has been hit the hardest and all three sectors fared similarly poorly concerning the amount of cargo transported.

4 Highly Volatile Rates

Comparing the income across all sectors with the sport freight rates to be obtained on container vessels, we see across observed data from almost 15 years a very low correlation, despite what we just saw for the amounts of cargo moved. First, we are taking the Clarksea Index which indicates the average earnings per day of the global fleet. It takes into account the average earnings the tanker sector (VLCCs, Suezmax and Aframax), of the dry bulk sector Capesize, Panamax, Handymax and Handysize), by some container vessels and gas tankers and is weighted by the

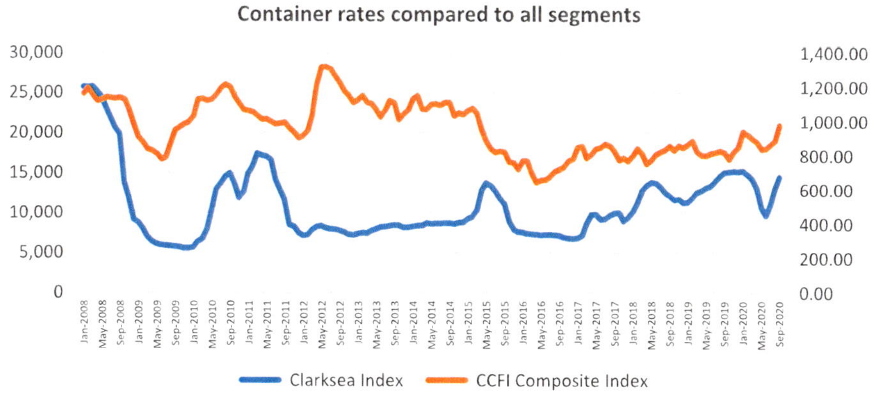

Fig. 5 Revenue in container spot rates as measured by the CCFI and revenue of all shipping segments. Source: Clarksons

number of ships in each sector. The index is matched with data from the aggregate container spot rates as expressed by the CCFI. From 2008 until the beginning of 2020 there is a weak correlation of around 0.2. In 2020 however, we see a very close correlation and both indicators move closely together with a correlation of almost 0.9 (Fig. 5).

We need to drill the observations further down into the container sector as the supply market is no homogenous. The container shipping segment consists of those that offer physical transport on vessels (operators) and those who provide the vessels (lessors or more colloquially owners). Both are commanding respectively rates for their services (freight and charter rates).

As we show in the following graph, freight and charter rates are clearly not moving in tandem when taking the last two decades to illustrate the case. Two indices help to understand the phenomenon: We take the aggregated CCFI as index to define the movements of spot freights rates. We neglect that spot rates only show part of the freight rates but as indicator this should suffice. We also neglect that is has been shown that carrier income per nautical mile varied even on front-haul service by a factor of four (alphaliner, 2020a, b, S. 1). Comparable data show that long-term freight rates also seem to strengthen at the end of 2020 (Fig. 6).[1]

Charter rates for container ships have been aggregated on a monthly basis by Clarksons and weighted for the various ship sizes. Both indices show strong volatility. In relatively normal or positive markets when growth occurs and in previous crisis both indices have moved rather independently. If correlations exited, they did so for very brief periods of time. COVID-19 has changed this occurrence in a rather unique way which deserves further attention. For the first time in the observed period of almost two decades both types of rates have moved almost in tandem, during the downward correction starting in late 2019 until the trough in May

[1]The Xeneta XSI index is showing an increase of 1.6% in November 2020, a y-o-y rise of 1.9%.

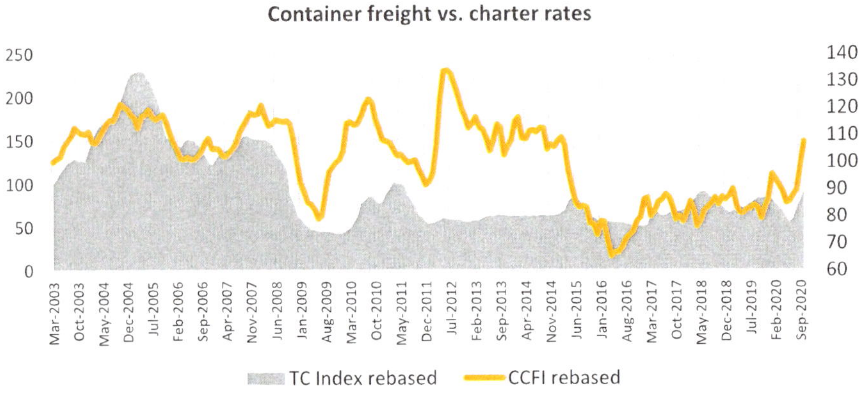

Fig. 6 Freight rates compared to charter rates (TC index, CCFI), both being rebased on 100 in March 2003. Source: Clarksons

Table 1 Annual correlation between freight and charter rates. Own calculation based on Clarksons

Year	Correlation freight/charter rates
2004	0.85501031
2005	0.44066923
2006	−0.65661234
2007	0.85013709
2008	0.82644465
2009	−0.17030758
2010	0.58562151
2011	0.8182509
2012	0.95084497
2013	−0.61766965
2014	−0.23774658
2015	0.1312832
2016	−0.38638027
2017	−0.5409082
2018	−0.57457517
2019	−0.44700935
2020	0.93870914

2020 and the strong upswing until the last recorded figures in October 2020. What is new here is the practical unique correlation between both movements. When container lines started to profit from fleet reductions and subsequently strengthened freight rates, it had an almost immediate follow-on effect for charter rates. While particularly in the critical year 2009 a negative correlation drove both parts of the market apart, the contrary can be observed in 2020. With are exceptions, never has the movement of freight and charter rates been so smoothly aligned (Table 1).

There appears a second paradigm shift when plotting both data sets together. Seven years of independent or even opposed movements of both rate types, 2020 appears as practically the strongest year ever in parallel directions, both into negative territory as well as the strong upswing in the middle of the global health crisis.

To understand possible causes of the new phenomenon, we will try to point out some data that could help explain why 2020 may mark a new trend that could shape the perception of the entire container shipping market for the decade to come. Firstly, it is instructive to note the situation of containerships in layup. Unusual since the beginning of 2019 have been the large number of containerships that have been taken out of service for several weeks at a time for scrubber retrofitting. On January 2020 new regulations on the global limit on sulphur in ships' fuel oil came into force. The global upper limit on the sulphur content of fuel oil was reduced to 0.50% (from 3.50%). Such regulation of the IMO ("IMO 2020") had a mandatory character for all ships operating outside designated Emission Control Areas (ECAs), where the limit was 0.10%.

The new limit as part of the International Convention for the Prevention of Pollution form ships (MARPOL) made it necessary for all ships to either bunker cleaner fuels or to install scrubbers. Such scrubbers are limiting air pollutants by installing exhaust gas cleaning systems. Scrubbers remove sulphur oxides from the ship's engine and boiler exhaust gases, enabling ships fitted with them to continue to use heavy fuel oil, in accordance with IMO Guidelines. A particularly large share of container ships was chosen to be fitted with scrubbers and thus had to reconfigured before or around the deadline of Jan 1st 2020. The TEU capacity of ships laying idle or in layup compared to those being retrofitted for scrubbers changed strongly between the middle of 2019 and Ende of 2020. While the idle ships were stable in the two last quarters of 2019. That figure almost tripled (in terms of capacity) towards to end of the second quarter 2020. Meanwhile, most retrofits were being installed in the last quarter of 2019 as could have been expected and decreased afterwards. That fact that still in Q3 and Q4 of 2020 some ships were being retrofitted with scrubbers had partly to do with the difficult situation on shipyards and dramatic delays—sometimes of several months—due to COVID-19. Importantly, it must be noticed that the idle capacity has been shrinking since the middle of 2020, despite the fact that less cargo was available. But operators were starting to charter in vessels at a rapid pace (Fig. 7).

Chartering in vessels at a time when many voyages were canceled seemed counterintuitive at first. Such simultaneous behavior of container lines could not be observed in previous periods of recovery. However, in the case of the year 2020 it coincided with a rapid rise in spot freight rates. The idling spiked in June 2020, the same time freight rates were taking up to a continuous rise until the end of the year.

Finally, it is important to note the occurrence of "blank sailings". Even though it is common that in low seasons or in troughs not all voyages of container vessels take place as scheduled by the shipping lines months in advance. Highly unusual was in 2020 to observe how blanked sailing became a matter of public announcements, discussions and were used as a tool to communicate the availability—or non-availability—of container slots. Capacity management is a under normal circumstances a daily tool for container lines to balance their fleet and adjust supply (of vessels and slots) to demand (of containers to be transported). Intuitively, a drop

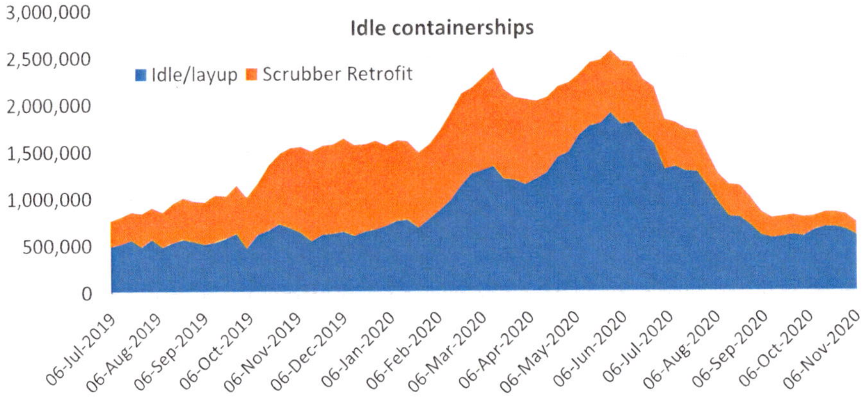

Fig. 7 Idle container ship capacity measured in TEU, differentiated for vessels undergoing retrofits with scrubbers and those being not deployed in idle state or in layup. Source: Clarksons

in container freight rates should be followed by a drop in capacity provided and vice versa. The higher the rates, the more capacity would be made available. In a simplistic microeconomic explanatory framework, the adjustment of slots and transport needs (supply and demand) should culminate in a relatively balanced price. It must be noted though, that container lines in the past have not been acting according to textbooks as entirely rational entities. A subtext for pricing strategies had been secondary goals of attainment of market share, dominance on certain routes and similar intentions. The market of container shipping lines had been divided among so many competitors that neither coordinated nor simultaneous behavior could not be observed, resulting in frequent underpricing of the transport service to the detriment of the entire sector.

Beyond frequently reported anecdotal evidence, the statistics of reported blank sailings in 2020 coincide counter-intuitively with the movement of freight rates. In particular in the second half of the year it becomes apparent that blank sailings have been (partly) used to support a raise in freight rates. Capacity has been shortened in order to keep up freight rates or increase them further. This observation is even more remarkable as these blank sailings have been distributed—more or less—evenly between all alliances and most independent operators. On the bottom line, the entire container shipping industry has benefited. This has been made possible by the absence market players that were seeking market share at the expense of profitability. One may go as far and state that an era of fight for share has been overcome and is being replaced by a phase of profit seeking (Fig. 8).

The graph illustrates the rational but—in the container shipping market—unusual behavior of shorting capacity in a rising market. The observation is steadily observable in the second half of the year. Such simultaneous behavior of an entire segment has raised the attention of several regulators. Notably the Chinese regulator voiced its discontent of high freights rates—fearing a dampening effect for reanimated exports—as well as the US regulator.

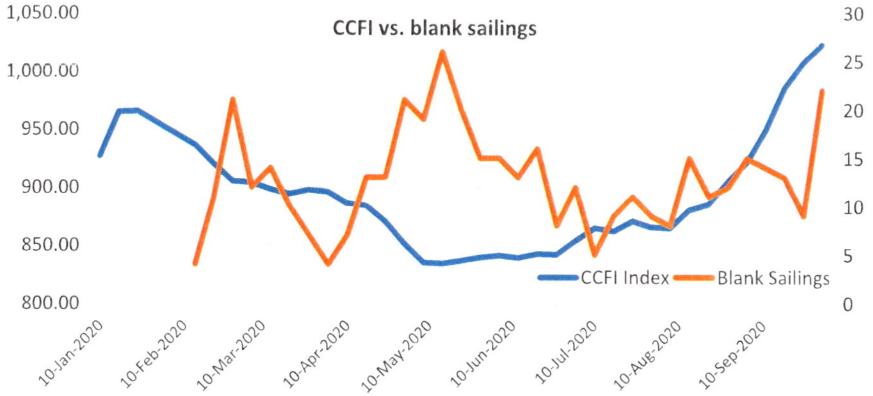

Fig. 8 Container spot rates (CCFI index) compared to blank sailings. The concurrent rise in blank sailings and freight rates is unusual. Sources: Clarksons, SeaIntellingence

The Chinese Ministry of Transport had invited 14 container lines active in transpacific trades to discuss their freight rates in early September. This meeting in the form of a "consultation" was supposed to exert pressure to dampen the increase of freight rates (Shen, 2020). Chinese container line COSCO did respond positively to the request, which other lines referred to market conditions. Chinese officials were particularly critical of blank sailings and active capacity management. The market intervention was criticized by non-Chinese container lines (Keohane, 2020).

In the US, the Federal Maritime Commission (FMC) announced to increase the level of its monitoring of container line alliances and requiring them to file carrier-specific trade data, on a monthly basis which was previously a quarterly basis. The FMC is particularly concerned with data for the Transpacific and Transatlantic relations and related rate hikes (FMC, 2020b). The FMC is also concerned with carrier alliance agreements and is trying to monitor their impact on shippers and the consumers for "adverse market effects". These measures followed concerns raised already in September, when the FMC admitted that it had difficulties to assess the highly unusual market situation, which it termed "unusual circumstances and challenges" as a consequence of the pandemic. The FMC acknowledged the receipt of "reports that addressed trends in spot rates, longer term service contracts, utilization of equipment, blanked sailings, revenue trends" (FMC, Commissioners Discuss Ocean Carrier issues, Global Alliances, and Market Developments, 2020a) that would be discussed.

While also Korean authorities questioned several container lines[2] on the rate setting in the transpacific market they were equally concerned that Korea may be left out in an overheated market in favor of more lucrative Chinese cargo.

[2]HMM, SM Line, Korea Marine Transport, Sinokor Merchant Marine, Maersk Line, MSC, CMA CGM, Yang Ming, Cosco.

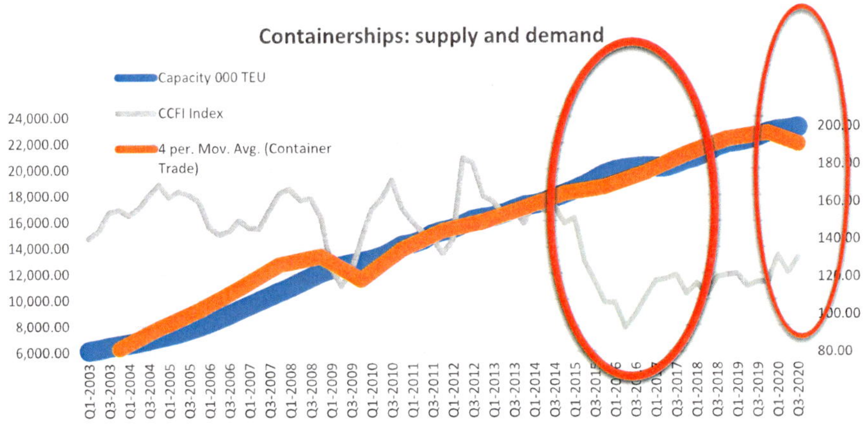

Fig. 9 Shown here is the imbalance between supply and demand of containerships with the resulting freight rates. Source: Clarksons

For the next dissection of the effects of blank sailings it is important to take into account the total available capacity in the container segment. This shows the price movements (of freight rates) that seem to be somewhat different to previous market constellations (Fig. 9).

The trendline of container trade shows the drastic drop in liftings as we had seen earlier. The only similar drop had been experienced in 2009 following the financial crisis. The trendline has been laid above the container capacity which shows as approximation the supply of container ships and their slots. The CCFI index as indicator of spot income per slot shows over most periods shown a reaction that would be expected. When demand was outpacing supply of container ships (most notable in the period 2004–2008) freight rates remained on a high level. When as suddenly as unexpectedly (black-swan-event) demand dropped below supply, rates tumbled. The regained strength in 2010 and followed the classical supply-demand pattern. The next remarkable situation occurred from 2015 onwards, when now supply outpaced demand and freight rated went onto a low level for longer period of time. The setback in freight rates put container lines under significant pressure. The industry could not sustain its structure of many small players. Since 2015 the low-rate period coincided with a fast-paced consolidation of container lines. The number of important market players was actively reduced, the largest ones gained in relative market share and the competitive basis was strengthened.

Turning the attention to the most unusual detail of observation we need to turn to the second and third quarter of 2020. While output in global production dropped markedly in the beginning of the year as a direct consequence of several countries in lock-downs, production reduction due to health considerations of workers and partly interrupted supply chains, the amount of cargo to be transported in containers and thus on container vessels dropped notably. Under normal circumstances this would have triggered a chain reaction with prices (i.e. freight rates) for the transportation service dropping. As can be seen in the above graph and documented through the CCFI index prices behaved in the opposite direction to the expected: upwards.

Firstly, this is giving us an indication that the supply side has behaved differently in 2020 than it did in 2008/09. Competitive pressure in the fight for market share could be avoided. Container lines could keep the focus on profitability.

Secondly, the rise in freight rates had a secondary effect which cannot be underestimated and is somewhat irritating for some market participants. Normally, shortening capacity by the container lines has a detrimental effect on charter owners. As most container lines have a fleet consisting of own and chartered tonnage, they have a natural tendency to reduce such chartered tonnage and keep owned tonnage in circulation. This would normally trigger a reaction of charter tonnage being handed back, increased idling of NOO tonnage (non-operation owners or lessors). This would have an immediate a negative effect on charter rates on the one hand and the length of contractual charter periods on the other hand. Following the positive rise in freight rates despite (or because) the shortening of supply had this time in 2020 only a relatively brief negative effect on charter rates. Almost immediately when freight rates could be stabilized charter rates could follow suit. This is remarkable because at least in the beginning not much additional tonnage was needed.

NOOs could markedly profit from the rise in sport rates with ships coming off charter in the end of 2020. With most lessors in private hands and little data being available, only publicly listed Seaspan (as part of Atlas Corp.) (Atlas reports Third Quarter 2020 Results, 2020) reported a substantial increase in net earnings in the third quarter. Similarly, Danaos (Danaos, 2020) and Global Ship Lease (GSL Global Ship Lease, 2020) announced increases of 25% in revenue due to the upswing.

5 Avenues of Explanation

There is no simple explanation for the phenomenon observed in the container shipping markets in 2020 triggered by the COVID-19 pandemic. We try to explore several avenues of explanations as possible patterns to understand observed data.

5.1 Impact of Consolidation

Container lines have shown a pattern of consolidation throughout the last decade. The background was repeated insights that larger lines tended to earn higher profits (McKinsey & Co, 2017). McKinsey suggested in its 2018 report also that mergers in the past had mostly resulted in revenue drops without strengthening the market position enough to keep prices on the same or higher level. Numerous mergers and takeovers have reduced the number of sizeable container lines that may offer transcontinental services (JOC, 2017):

Maersk:	Safmarine; Sealand (1999); P&O Nedlloyd (2005), Hamburg Süd (2017), Mercosul (2017)
Hamburg Süd:	SPCL (1999), Ellerman, Kien Hung (2003), Ybarra (2005), FESCO (2006), Costa Container (2007), CCNI (2015)
CMA CGM:	McAndrews (2002), Delmas (2006), U.S. Lines, CoMaNav (2007), NOL/APL (2016)
COSCO:	China Shipping (2016), OOCL (2016)
ONE:	NYK + Mitsui + K-Line
Hapag Lloyd:	CP Ships (2005), CSAV (2014), UASC (2017)
PIL:	Marina Express (2015)
Matson:	Horizon (2015)

It is thus safe to assume that a more consolidated market place may have an observable impact on consequential pricing, pricing strategies and adherence to such strategies.

UNCTAD has more recently investigated the concentration of container shipping lines on the important trans-pacific routes (UNCTAD, 2019, S. 46). They used data from 2006 (thus before the financial crisis during a booming market) and compared them with those from 2019, shortly before the outbreak of COVID-19. UNCTAD employed five different indicators for the investigation: The share of top shipping company (in %), the share of top four shipping companies (in %), the Herfindahl–Hirschman Index, the absolute number of companies active and the Gini coefficient. On all indicators except for the absolute number of companies the concentration in a statistical relevant manner. The top container line increases its market share by four percentage point (29–33%), the four largest broke the threshold of 60% (from 57%), the Herfindahl-Hirschman Index climbed by almost 20% and the Gini coefficient increased by more than 10%. Similarly, OECD had detected and illustrated the same effect in 2017 (OECD, 2017, S. 85) for the global scale. Other research had indicated the same trend earlier (Luo et al., 2014). Using these data and continuing the timeline (alphaliner, 2020a, b) the picture becomes clear: container operators have been on an accelerated path of consolidation in the last half decade. Looking back over two decades, the market share increased from 39% to 86% for the top 10 lines and from 24% to 59% for the grouping of the top 4 lines. Both have more than doubled. They increased by 120% and 150% respectively (Fig. 10).

The steady trend towards further consolidation has led several authors to question if the market of container shipping lines is still competitive or may should tendencies of negative concentration effects (Charłampowicz, 2018; Lee & Song, 2017). Charlampowicz shows that the Herfindal-Hirshman index for the top 10 container shipping lines has almost tripled in the last decade.

Higher concentration and consolidation is viewed critically across all markets. In particular the liner shipping industry has a long history of critical regulation namely from the EU Commission and other trade authorities. It is suspected that joint price setting and price cartels would harm the consumer. This has been most notable in the Commission Regulation (EC) No 906/2009 on the application of Article 81(3) of the

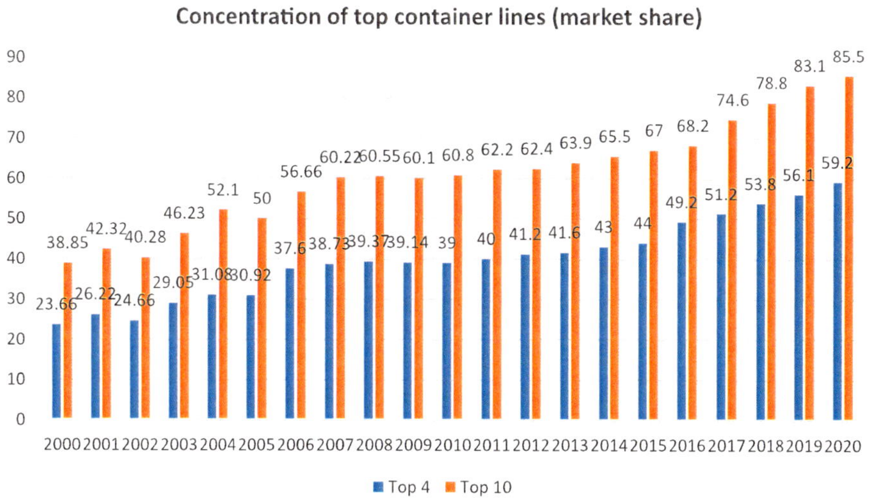

Fig. 10 Increasing consolidation of container lines is leading to measurable concentration. Source: OECD, alphaliner, Lloyds List

EU Treaty to certain categories of agreements, decisions and concerted practices between liner shipping companies (consortia). While the respective regulations had been modernized in 2009, they have been regularly extended, the last time in March 2020 (EU Commission, 2020). This "Liner Consortia Block Exemption Regulation" as it is known in the industry is based on the observation that a too fragmented industry would have a negative impact on pricing and the competitive environment.

Taking the above shown data into account it appears that the two observations are important. Firstly, the container liner shipping industry seems to have encountered the second black-swan-event in 2020 better, faster and much more proactive than the previous one. Secondly, it appears that contrary to previous assumptions the charter owners as supply side of the transportation service supply side have been towed out of the trough immediately after freight rates were spiking despite no immediate need for additional tonnage. Faster market movements and increased volatility may prove to be a positive fact. However, this applies to the spot market for charter rates. Notwithstanding to this observation is the market power of consolidated container lines in the negotiation of long-term charter contracts.

5.2 *Impact of Digitalization*

Pandemics, wars and other large scale disruptions to maritime trade are not new and have influenced the shipping markets frequently and profoundly. The particular interest in black swans in the twenty-first century is also due to the fact that information technology has made leaps to provide practically real-time data on vessels, cargo, shifting interests and demand. This opens the possibility to adjust supply and prices in real time, too. Hence an important focus on the various impact levels of digitalization (Saxon & Stone, 2017), which in this respect has less to do with the wide ranging effects on the operational part of vessels (Johns, 2018) but more with the effects on the shore side operation.

As we had seen above, the reaction curve seems to have changed in terms of speed over the observed quarters after the first sightings of both black swans, to remain in the picture. Comparing the black swans of 2008 and 2020, the largest paradigm shift that has occurred in industries across the board is certainly the influence and disruptive nature of digitalization. Digital technologies have taken many strands of business life and communication. In the maritime industry it has allowed to deploy resources in an even speedier fashion. In container shipping, liner companies were well adjusted for the fastest and unlikeliest kind of events. Despite the black swan type of event liner companies for themselves and the industry at large was able to adjust quickly. This applied to volumes and prices alike, resulting in a creative destruction of expectations. Instead of reacting with lower prices to an environment of a sudden drop in volumes, the situation gave rise to what felt like a large-scale industrial experiment. Supply of ships and slots was shortened even more pointedly than demand had dropped. Consequently, prices for the remaining goods have been rising quickly and strongly. Playing with capacity to raise prices needs three ingredients: a certain concentration in the market as shown above, trust in the inherent low elasticity of demand confronting increasing prices and global distribution of complex real time data as precisely made possible through a higher degree of digitalization of shipping companies.

5.3 *Impact of a Generational Learning Curve*

It has often been said that the shipping industry and in particular the container shipping industry is not learning from its mistakes. The data as presented here seem to draw a first sketch of a different picture, though. It seems especially container shipping lines have learnt a hard and commercially painful lesson. The fight for market share has been a costly exercise while profits may have been neglected. At the beginning of the pandemic it seemed a real and likely danger that the container operators would incur losses at a large scale while GDP seemed in free fall. This imminent danger has been avoided and turned into opportunity with container lines reporting at the end of the third quarter 2020 some of the highest profits in years.

An ingredient of this different attitude and approach may have to do with the fact that a different generation of managers is in charge of the profitability of liner shipping companies, of which only few are stock listed (Maersk, Hapag-Lloyd, COSCO, Evergreen, PIL) and responding to public shareholder interests. It is a different mindset that is putting profitability above market share, volume and leadership. Xeneta CEO Patrik Berglund captured this unexpected rational behavior the following way: "It is a strategy that continues to pay dividends. Spot rates remain high across the board, with capacity being managed expertly, much to the humiliation of many within the shipper community." (Xeneta, 2020) Part of story may also be that in the course of numerous mergers container lines have become less national. While 20 years ago it appeared as if container lines were almost regarded as national flag carriers (with the implicit obligation to fly the respective national flag on its vessels), large container lines are now multinational by definition. Another expression of this internationalization is the composition of managing boards of container lines.

5.4 *Impact of a Split GDP*

As indicated previously, global GDP or the breakdown of national GDPs may be a helpful indicator of merchant trade and eventually seaborne trade as a derivative. However, in particular situations physical production and provision of services may not grow and shrink by the same rates and react to the same stimuli. The global pandemic has shown that numerous services have undergone stoppages, difficulties and reductions. The travel and airline industries are just some examples together with restaurants, the entertainment industry and sports. While the decline in the first phase of the pandemic appeared to occur at similar speeds, restrictions in the second and third waves were much more differentiated. Production (as exemplified by the automotive industry) continued almost unhindered, while the service industries suffered repeated setbacks. Even more, household budgets were newly distributed with less expenses allocated to services and more to physical acquisitions. For shipping within the transportation industry as service sector with a derived demand this supported its rapid recovery. In the future it is suggested to alternatively focus less onto GDP as an early indication of possible seaborne trade but rather onto the primary (agriculture) and secondary (industrial production) parts of GDP only and neglect the service part. This would make respective predictions more complex and may defy previous common approaches but the 2020 COVID-19 pandemic gave important evidence that the broad-based GDP is an insufficient explanatory mean for a predictive analysis of seaborne trade.

5.5 *Playing Prices, Not Following Them*

We have mentioned already that container shipping lines are measured against their degree of concentration. Clearly, concentration has increased in the last decades among liner companies while a similar mirroring effect could not yet be observed for the charter owners (lessors) of container vessels.

It will be important to study carefully if and when this highly competitive market pivots towards an oligopolistic market with negative effects on efficiency and pricing. As mentioned above, competition authorities in Asia, the US and Europe have indicated their vigilance in this respect and container liner shipping has a long track record of competitive scrutiny. However, it cannot naively be assumed that fewer companies lead necessarily to less fierce competitive behavior. The beneficial arrangement of container shipping alliances had already indicated in more fragmented markets that a certain weight from market players may being a healthy correction to pure market forces. Observations from 2020 now suggest that the liner part of the industry has transformed itself to a higher level of efficiency. If this argument can be supported the economy would benefit. Immediate beneficiaries of this situation have been charter owners as shown above.

5.6 *New Pricing Models*

This paper does not investigate in more depth the impact of new pricing structures that all larger container lines had introduced with the advent of new global sulphur regulations in the beginning of 2020 that effects of which have been overshadowed by the pandemic. Container lines had to find ways to transparently share the additional costs for cleaner and more expensive fuel with their customers. Hapag-Lloyd had introduced its marine fuel recovery (MFR) plan with all customers and used the ITC Transition Charge (Hapag Lloyd, 2019) for the initial transitory phase. Maersk remodeled the bunker adjustment factors.

5.7 *Understanding Elasticities*

The share of ocean freight in the entire cost tableau for almost all goods is low, and can oftentimes be practically neglected. This is a strong indicator for the low elasticities of freight rates. The more consolidates the industry is the more likely it becomes that confidence in market power is increasing. Price elasticities can only actively be played in a market where few players have a larger market share. They would act rationally (and not collude) if they understood and tested the possibilities to be active managers of prices and not merely price takers. If this kind of behavior can be replicated in future market situations it would open up a new field of pricing strategies for container liner companies. Thus far, the fragmented market allowed often only special and differentiated pricing strategies with niche service such as reefer cargo and similar well-defined offers.

5.8 *Impact of Orderbook*

A further avenue which we have not explored in this paper and which would deserve further attention is the dynamic impact of the respective order books before, during and after the two black-swan-events that are observed here. While the container order book at the end of 2007 was standing at a historic high with 61.25% of the fleet in the water on order, the contrary can practically be said for the situation in early 2020. Here the orderbook in the container segment stood around 10% of the existing fleet, the lowest figure on record.

The run up to 2007 had been produced buoyant markets over several years, an advantageous financing environment especially in the German market that brought the largest share of container shipping new buildings and an optimistic environment that assumed continued linear growth patterns. The run up to 2020 had seen much more cautious markets. Even though money was cheaper than in most previous decades and amply available, neither operators nor lessor showed an appetite to invest in additional fleet. The regulatory context before 2020 had seen the implementation of global sulphur rules and the threat of additional Emission Control Areas (ECAs) as well as large scale discourse on climate friendly additional regulation. As notably the climate debate circled around CO2-reduction measures it was unclear if global or regional regulations would move first and influence design, costs structures and fuel types of vessels more and thus creating significant insecurity for investment cases that needed to hold over more than 20 years.

Both figures would certainly exercise a significant pressure on the market situation with an impactful delivery schedule post 2008 and the opposite in 2020.

6 Conclusion

We have undertaken in this paper to investigate the behavior of container shipping lines in the COVID-19 pandemic. We have used the topos of the black swan to illustrate that we observed a non-standard situation. What could be observed were striking differences in reactions, when comparing the 2008/09 financial and economic crisis and the 2020 global pandemic.

We have seen a rather differently assorted group of container lines on a significantly higher level of consolidation using their situation in a different manner. The container lines have shown in 2020 less of a behavior of price takers, they seemed to have been able—at least partly—to set prices through shortening of supply against usual reaction patterns. This has led to price (freight rate) increases that have more than compensated for the initially very high drop in volumes as measured by boxes transported. The adaption of supply was expressed by so-called "blank sailings".

A further observation has shown that the profitability of the operator side of the container market has helped the secondary players (container ship owners or lessors) have been drawn out of the trough faster than expected.

This leads to the counter intuitive idea that the increase in consolidation may have—at least some—positive effect also for the tonnage providers which would normally be more skeptical about the increased market power of their main customers. It will be worth observing closely if and how such market power will play out in the long term and if the new constellation may have carved out a new niche for the tonnage providers or if it would rather lay the groundwork for a disruptive development.

The observations presented have narrow limitations which need to be carefully studied further. Not included in the data has been the impact for non-comparable fuel prices in both black swan periods of crisis. Previous research would need to be updated (Beverelli, 2010). Equally not included is the possible impact of the vastly different levels orderbooks on short and long-term behavior.

References

alphaliner. (2020a, November 28). *Alphaliner TOP 100*. Retrieved from https://alphaliner.axsmarine.com/PublicTop100/

alphaliner. (2020b, November 04). Box shortage pushes up rates on headhaul trades and starts to. *alphaliner, 45*.

Atlas reports Third Quarter 2020 Results. (2020, November 09). Retrieved from https://ir.atlascorporation.com/press-releases

Avishai, B. (2020, April 21). The pandemic isn't a Black Swan but a portent of a more fragile global system. *The New Yorker*. New York.

Beverelli, C. (2010). *Oil prices and maritime freight rates: An empirical investigation*. UNCTAD.

Charłampowicz, J. (2018). Analysis of the market concentration of the container shipping markets–selected issues. *SHS Web of Conferences 58, 01005 (2018)* (p. 6). GLOBMAR 2018.

Council, P. (2020, September). Lessons of past coronavirus pandemics. *Population and Development Review*, 633–637. Retrieved from https://doi.org/10.1111/padr.12360

Danaos. (2020, November 05). *Danaos corporation reports third quarter and nine months results for the period ended September 30, 2020*. Retrieved from https://www.danaos.com/investors/press-releases/press-release-details/2020/Danaos-Corporation-Reports-Third-Quarter-and-Nine-Months-Results-for-the-Period-Ended-September-30-2020/default.aspx

EU Commission. (2020, March 24). *Antitrust: Commission prolongs the validity of block exemption for liner shipping consortia*. Retrieved from https://ec.europa.eu/commission/presscorner/detail/en/ip_20_518

FMC, F. M. (2020a, September 16). *Commissioners Discuss Ocean Carrier issues, Global Alliances, and Market Developments*. Retrieved from https://www.fmc.gov/commissioners-discuss-ocean-carrier-issues-global-alliances-and-market-developments/

FMC, F. M. (2020b, November 25). *Federal Maritime Commission Increases Global Alliances' Information Monitoring Report Requirements*. Retrieved from https://www.fmc.gov/federal-maritime-commission-increases-global-alliances-information-monitoring-report-requirements/

GSL Global Ship Lease. (2020, November 09). *Third Quatrter 2020 - Results Presentation*. Retrieved from https://www.globalshiplease.com/static-files/f4da65b5-24af-4ac0-bd7d-fd654f686a96

Hapag Lloyd. (2019, November 28). *Worldwide Price Announcement – IMO 2020 Transition Charge (ITC)*. Retrieved from https://www.hapag-lloyd.com/en/news-insights/news/2019/10/worldwide-price-announcement%2D%2D-imo-2020-transition-charge%2D%2Ditc-.html

JOC. (2017, July 16). *Joc.com*. Retrieved from Container Carrier Mergers & Acquisitions, 1977 to present: https://www.joc.com/maritime-news/container-lines/container-carrier-mergers-acquisitions.html

Johns, M. (2018, October). *Seafarers and Digital Disruption. The effect of autonomous ships onthe work at sea, the role of seafarers and the shipping industry*. London: ICS International Chamber of Shipping. Retrieved from https://www.ics-shipping.org/wp-content/uploads/2020/08/ics-study-on-seafarers-and-digital-disruption-min.pdf

Keohane, D. (2020, November 17). China presses shipping line to rein in record freight rates. *Financial Times*.

Lee, C.-Y., & Song, D.-P. (2017). Ocean container transport in global supply chains: Overview and research opportunities. *Transportation Research Part B, 33*.

Luo, M., Fan, L., & Wilson, W. W. (2014, January). Firm growth and market concentration in liner shipping. *Journal of Transport Economics and Policy*, 17.

McKinsey & Co. (2017). *Shipping M&A destroying shareholder value*. McKinsey.

OECD. (2017). Market concentration of container shipping lines 2000-16: Market share of container lines (%). In *ITF Transport Outlook 2017*. Paris: OECD Publishing.

Saxon, S., & Stone, M. (2017). *How container shipping could reinvent itself for the digital age*. McKinsey. Retrieved from https://www.mckinsey.com/industries/travel-logistics-and-transport-infrastructure/our-insights/how-container-shipping-could-reinvent-itself-for-the-digital-age

Shen, C. (2020, September 13). Carriers summoned by China to curb transpacific mark-ups. *Lloyds List*. Retrieved from https://lloydslist.maritimeintelligence.informa.com/LL1133882/Carriers-summoned-by-China-to-curb-transpacific-mark-ups

Shermer, M. (2014). How the suvivor bias distorts reality. *Scientific American*.

Smith, G. (2014). *Standard deviations: Flawed assumptions tortured data and other ways to lie with statistics*. Overlook.

Taleb, N. N. (2007). *The Black Swan. The impact of teh highly improbable*. Random House.

Taleb, N. N., & Spitznagel, M. (2020, March 31). Die Corona-Pandemie ist kein schwarzer Schwan. *NZZ Neue Zürcher Zeitung*, 29.

UNCTAD. (2019). *Review of Maritime Transport 2019*. UNCTAD.

Xeneta. (2020, November 25). Retrieved from Xeneta: https://www.xeneta.com/news/container-rates-alert-no-crisis-for-carriers-as-long-term-contracted-rates-and-earnings-continue-upward-trajectory-but-a-halt-may-come-in-early-2021

COVID-19: Challenges and Future Responses in International Logistics

Sung-Jun Park and Eun-Soo Kim

Abstract The main purpose of this chapter is to provide some understandings and key words hopefully useful for building a new framework in the arena of international logistics and its relevant industrial sectors, responding to the uncertainty and challenges mainly generated by the outbreak of COVID-19 pandemic, an unprecedented and probably the most disruptive global supply chain risk event since the 2000s to the present. The Chapter consists of three parts: literature review focusing on the pandemic as a supply chain risk (SCR) source; revisiting the impacts of COVID-19 on global economy and international logistics; a preliminary exploration of a new approach to SCR mitigation. Through a brief literature review, this chapter argues that SCR mitigation strategy specifically focused on impacts of the pandemic as a SC disruptor in global scale has been relatively underdeveloped. By providing some descriptions of the COVID-19 impacts on GSC, companies' responses in global market, and experts' observations and analyses, authors attempt to distill the differences of nature and mechanism of SC disruption between pre-COVID-19 and post-COVID-19, which can be characterized by combination of supply shocks and demand shocks, and rapid move of business operations to the "contactless" world, replacing at least partly efficiency with resilience as priority, while human factor playing a most unique and formidable role especially when encountering the SC disruption. In order to find bottom line for building a new framework for SC mitigation strategy specific to pandemic-caused SC disruption, the authors, in third part of the chapter, introduce "key words" recently re-emerged and/or discussed in relevant research arena, which include preparedness, resilience and agility, digitalization, and SC reconfiguration. In conclusive part ("Implications for international logistics"), the authors urge that all of key words require great attentions from international business world as well as serious consideration from governments for navigating this changing world filled with growing uncertainty.

S.-J. Park (✉) · E.-S. Kim
Korea Maritime Institute, Busan, Republic of Korea
e-mail: sjpark@kmi.re.k

B.-W. Ko, D.-W. Song (eds.), *New Maritime Business*, WMU Studies in Maritime Affairs 10, https://doi.org/10.1007/978-3-030-78957-2_4

1 Introduction

Since its outbreak in the early January 2020, the COVID-19 pandemic has rapidly moved global concerns over treating infectious disease from the issue of keeping global human health to the issue of global economic health intertwined with global supply chain (GSC) disruption and supply chain risk management (SCRM), which is directly associated with future course of international logistics as well as global economy.

In general, before the outbreak of COVID-19 and even during its initial take-off period of rapid spread, significance of infectious disease as a major source of GSC disruption has been broadly neglected, especially in comparison with other natural disasters such as tsunamis and earthquakes in terms of GSC risk sources. In the early stage of COVID-19 crisis, there were few who had imagined that the impact over the world economy should be the worst to next to "the Great Depression" in the 1930s. As time went by, however, such a scenario has become an undeniable reality that people has never seen as follows:

The death toll of the COVID-19 has already approached to be equal to the level resulted from the major wars experienced in the twentieth century.[1] On the other hand, world economy has also experienced the "deepest global recession since the Second World War," which can be summarized by the "fastest, steepest downgrades of growth among all global recessions since 1990" (World Bank, 2020).[2] Despite a certain level of improvement, the growth rate of world economy in 2020 resulted in −4.2% (OECD, 2020),[3] which proved the worst that had been expected earlier in the same year.

The COVID-19 dealt straight and successive blows on world trade more severely than on the world economy. World merchandise trade volume in 2020 is expected to fall by −9.2%, which is the worst record since 2015 (WTO, 2020), and even overshadows the "Great Trade Collapse" in 2008–2009 (Baldwin & Evenett, 2020).

Against this backdrop of gloomy observations, everybody began to talk over a "new normal," which can be characterized as follows:

Firstly, a way of life. So called the concept of "un-tact" (meaning "contactless" in Koreanized English) gathers strength as a new but normal way of life at every corner of the world. A series of consolidated social distancing measures for responding to and curbing the disease—in parallel with waves of the crisis—accelerated the un-tact aspect of life to be more normal rather than abnormal or exceptional.

---

[1]The official death toll in the Korean War (1950–1953) is nearly 5 million. The death toll resulted from the COVID-19 for 11 months in 2020 approached 2 million (https://www.worldometers.info/coronavirus/coronavirus-death-toll/).

[2]World Bank, "2020 Year in Review: The impact of COVID-19 in 12 charts," (https://blogs.worldbank.org/voices/2020-year-review-impact-covid-19-12-charts).

[3]See OECD (http://www.oecd.org/economic-outlook/december-2020/). The growth rate is tentative for January 10th 2021.

Secondly, a way of thinking. A myth of globalization, which had, for several decades, been a leading way of thinking for global trade and international business, a strong stimulant for off-shoring strategies and global integration of production networks and supply chains was broken down, being replaced by a rising question of its vulnerability to SC disruption—highlighting the issue of "de-globalization."

In this changing world –in terms of way of life as well as way of thinking—in every aspect including production, consumption, international trade, and GSCs, questions of how to address and/or how to adapt to the COVID-19/post COVID-19 world raises a big question to international logistics industry not only to householders, companies, but to governments.

In this chapter, authors will provides some clues to solve the question—mainly by reviewing of existing research endeavors, finding lessons with a rough sketch of future prospects.

2 Pandemic and Global Supply Chain

2.1 A Brief Literature Reviews on Supply Chain Risks

2.1.1 Major Researches Before COVID-19

Jutter et al. (2003) conducted a field research to clarify the concept of supply chain risk management (SCRM), which contains interview findings of practitioners such as manufacturing companies, retail companies, and logistics service providers—in the wake of global "millennium bug" (Y2K) confusion.

The authors laid a foundation for future research of SCRM by providing the four basic constructs: assessing the risk sources for supply chain; identifying the risk concept of the supply chain by defining the most relevant risk consequences; tracing the risk drivers; and mitigating risks in the supply chain. In this research, the authors divides the risk sources into three categories (environmental, organizational, and network-related), in which of the environmental risk sources contains accidents, socio-political actions and "acts of God" including extreme weather or earthquakes.

Veselko and Bratkovic (2009) focus on globalization process and the information technology as major dynamic factors amplifying "security problem" (or "risks and threats") in international logistics chains. While introducing the "risks" provided by existing researchers, the authors attempt to reshuffle the "threats" into three levels: application level (natural disasters, illegal acts, low confidential data security etc.), organizational level (legislation, strategic decision making, lack of investment etc.), and inter-organizational level (for example, weak and inefficient company control over the operations of its suppliers, customers and the entire logistics chains). One figure that deserves highlighting in the research is that the authors include "the epidemics" as one of the threats in the application level. Despite that, further research has yet to be conducted for emergence of the issue of the "pandemic."

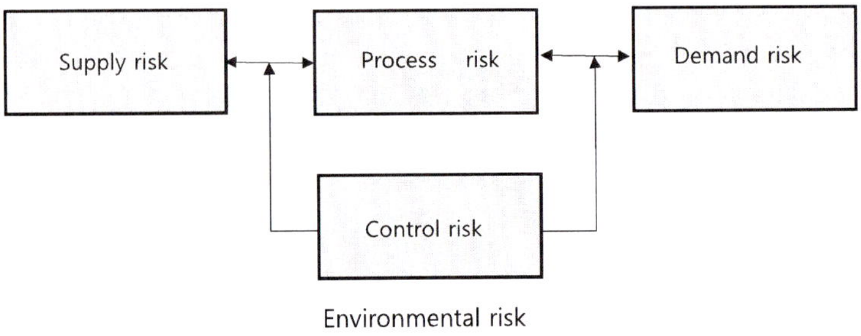

Fig. 1 Sources of risk in the supply chain. *Source:* Christopher (2011)

In the 4th edition of his well-known publication, Christopher (2011) puts the emphasis on growing vulnerability of global supply chain led by advancement of globalization and increasing out-sourcing phenomenon, which means, as the author calls, the "extended of supply chain." Synthesizing earlier discussions, the author provides five sources of supply chain risk, which include: supply risk, demand risk, process risk, control risk, and environmental risk (see Fig. 1). The author did not mention the pandemic issue, but he laid a foundation for reminding SC audience of adverse effect from the extended supply chains especially originating from China (then "world factory"), and provided a momentum for discussing "resilience"[4] as an option to respond to earlier questions in the logistics context.

At a time of increasingly growing concerns on the SC disruption, World Economic Forum (2012) reorganizes SC risks as "triggers of global supply chain disruptions, which are divided into 4 categories/19 triggers in an agenda-setting report (see Fig. 2). In this report reflecting a survey, the "pandemic" is included as a trigger with natural disasters and extreme weather in the "Environmental" category. Interestingly, 11% of all the respondents in the report think that the pandemic is "influenceable," which shows the level of awareness of logistics practitioners at the time.

Sodhi and Tang (2012) coin the "Butterfly Depiction of SC risk" for preventing causes (before the risk event) and for responding impacts (after the occurrence of the event), while reorganizing SC risks as two types—disruption and delay. In their work, risk drivers that create disruption are represented by natural disasters, labor disputes, supplier bankruptcy, fires and terrorism, and "dependency on a single source of supply as well as capacity and responsiveness of alternative suppliers."[5] The authors contribute to planning of mitigation strategies by conceptualizing of the tools such as alignment, flexibility and buffering, but still pays little attention on the issue of the pandemic.

---

[4]In his book, the resilience was defined as "the ability of a system to return to its original or desired state after being disturbed. See Christopher (2011), pp. 206–207. But the "resilience" was introduced by Sheffi in his book, *The resilient enterprise* published in 2005.

[5]For detailed description, see Sodhi and Tang, *Managing Supply Chain*, p. 55(Table 4.1).

Category	Triggers	rate	Level of control
Environmental	Natural disasters	59%	Uncontrollable
	Extreme weather	30%	Uncontrollable
	Pandemic	11%	Influenceable
Geopolitical	Conflict and political unrest	46%	Influenceable
	Export/import restrictions	33%	Controllable
	Terrorism	32%	Influenceable
	Corruption	17%	Influenceable
	Illicit trade and organized crime	15%	Influenceable
	Maritime piracy	9%	Influenceable
	Nuclear/biological/chemical weapons	6%	Influenceable
Economic	Sudden demand shocks	44%	Influenceable
	Extreme volatility in commodity prices	30%	Influenceable
	Border delays	26%	Influenceable
	Currency fluctuations	26%	Influenceable
	Global energy shortages	19%	Influenceable
	Ownership/investment restrictions	17%	Controllable
	Shortage of labour	17%	Influenceable
Technical	Information and communication disruptions	30%	Controllable
	Transport infrastructure failures	6%	Controllable
			■ Uncontrollable ■ Influenceable ■ Controllable

Fig. 2 Triggers of global supply chain disruptions. *Source:* WEF (2012)

Jonas (2012) makes a valuable contribution toward developing the SCRM, as an external expert working in her position of Avian and Human Influenza Response Coordinator in World Bank—by reminding characteristics of pandemic risk, by connecting the issue with development issue and by suggesting response strategies from the perspective of global risk-management, narrowing a gap between health risks and SC risks. In a paper written with pandemics of SARS (2003) and H1N1 (2009), she warns "each occurrence tends to be treated as a one-off surprise as news with an outbreak narrative,"[6] emphasizing the "economic impacts" such as systemic economic shock, and reminding the importance of systemic resilience, most of which reappears in COVID-19 pandemic in a more profound manner. Especially she warns the disruption of SC in pandemic has cascading failure in various areas at public sectors (security, health, education) as well as private sectors (food, water, power, transport, banking, communications, etc.).[7]

One point worth a special note argued in her report is that interconnectedness—which means dependence on long supply chains, just-in-time deliveries, and

[6] See Jonas (2012), p. 12.

[7] See Jonas (2012), p. 3.

growing of global production networks—may expose global economy to vulnerability of pandemic impacts.

2.1.2 Findings

Experts' concern on the SCRM of pandemic and its treatments as research agenda started in early 2000s in tandem with advancement and intensification of globalization. It is noteworthy that, in first 10 years of twenty-first century, experts' discussion focused on globalization as gradual increase of vulnerability, which is closely related with the impact of the pandemic.

In general, however, the issue of correlation between pandemic and global SC disruption seems not to be sufficient due to the reasons, as Jonas put it in her paper, that "each occurrence tends to be treated as a one-off surprise as news with an outbreak." Significance of the pandemic issue in SCRM originates from the fact that its impact and/or mitigation strategy are depending on little known tools—for example, avoidance reactions and social distancing that far beyond the sphere of logistics practitioners.

2.2 *Revisiting COVID-19*

2.2.1 Impacts on Global Economy and Trade

The global economic impacts or "shocks" of COVID-19 pandemic began to be glaringly sensed in March–April 2020 when the world had suffered over three million confirmed cases, and for curbing the disease, and governments around the world have imposed draconian containment policies-for example, lockdowns and even export restrictions (Baldwin and Evenett, 2020). It is very around this time that experts around the world began their efforts to analyze its ways of impacts, to measure scale of the impacts, and to come up with response strategies.

In their e-book, Baldwin and Weder di Mauro (2020) reorganize the economic shocks of the COVID-19 into three types—purely medical shocks, economic impacts from containment measures, and the expectation shocks. First, the medical shocks means that patients in bed and potential in social distancing could not participate fully or partly production of GDP. Second, the containment measures have also economic impacts, which was visibly shown, for example, in Hyundai Motors' case that its Ulsan factory located in South Korea could not avoid a temporary shutdown by supply disruption of parts and components from China—due to Chinese government's domestic lockdown measure.[8] Third and finally, the

[8]Kookjae dailynew, Feb 17, 2020 (in Korean).

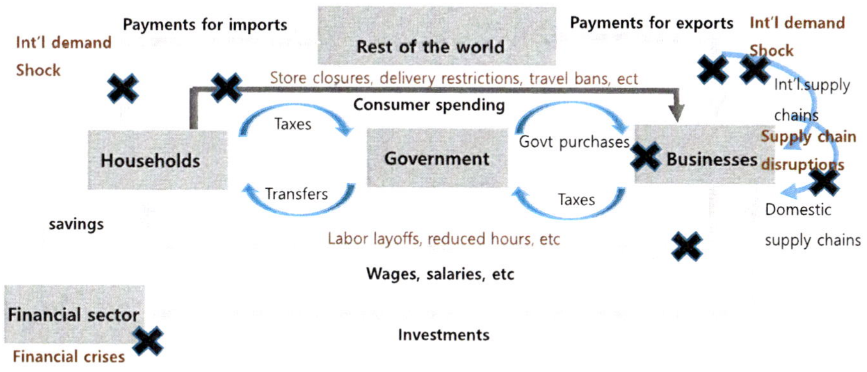

Fig. 3 COVID-19 economic impacts. *Source*: Baldwin and Weder di Mauro (2020)

expectation is closely related with negative effect of typical behaviors in economic crisis and uncertainty—for example, restraints or distortion of spending.

The most dreadful trait in the spread of COVID-19 crisis is that the economic impacts from the pandemic explode in combined mixture of demand shocks and supply shocks, having strong cascading effects, which originates from the interconnectedness (as mentioned by Jonas) or "complex web of interconnected parties" (Baldwin & Weder di Mauro, 2020, p. 10; pp. 44–45), as shown in a circular flow of diagram (Fig. 3).

As a comprehensive result of complex combination of demand shocks and supply shocks (supply chain disruptions), the world economy has experienced considerable shrinks. World GDP growth in 2020 was expected to be −4.2% growth (OECD, 2020).[9] Global trade volumes contracted sharply in the first half of 2020, with merchandise trade falling by 16% from its pre-pandemic level, especially dealing blows on international travel and tourism (Fig. 4).

Despite signs of slow recovery, world trade in 2021, as a downside scenario runs, is expected to suffer from the drop in demand, while declining trade growth—mainly because of a longer period of weak demand, weaker risk appetite and repricing in financial market (OECD, 2020). According to the WTO's projection, global trade volume growth should rebound from −9.2% in 2020 to +7.2% in 2021, which would remain below pre-COVID-19 crisis (WTO, 2020).

2.2.2 Impacts on Global Supply Chain and Maritime Industry

Rodrigue (2020) displays how the COVID-19 has impacts on SCs by triple waves of shocks such as supply shocks, demand shocks, and distribution constraints. The supply shocks means that the outbreak of the pandemic would occur an unexpected

[9]The WTO's projection of GDP growth is −4.8%. see WTO press/862 press release, 6 October 2020 (https://www.wto.org/english/news_e/pres20_e/pr862_e.htm).

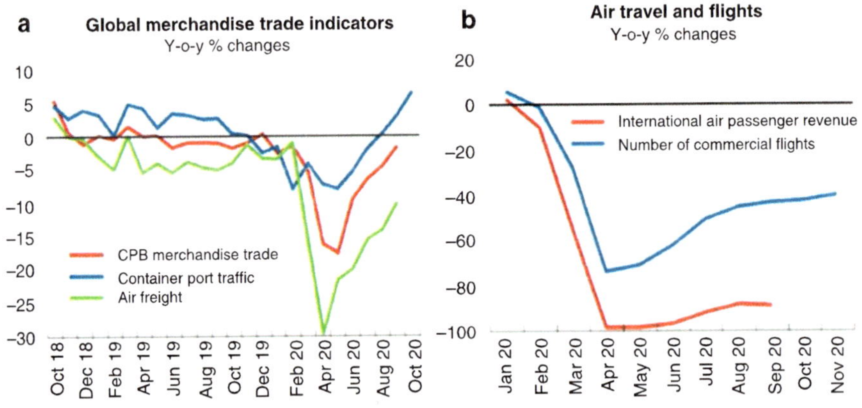

Fig. 4 Global trade and travel/flight growth rate in 2020. *Source*: OECD (2020)

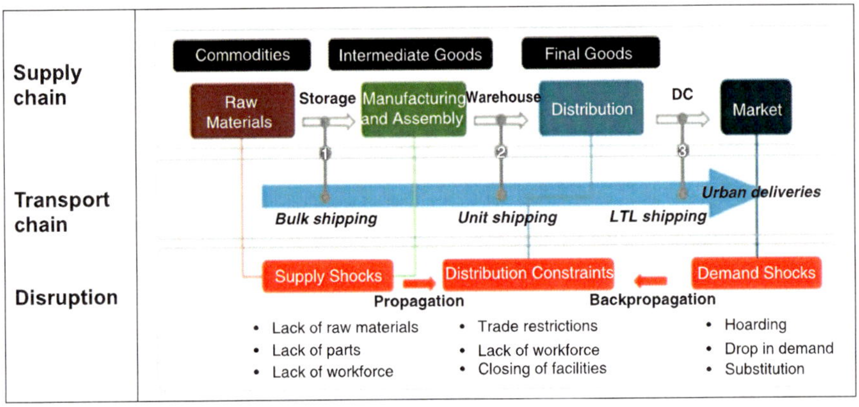

Fig. 5 Impacts of pandemic on supply chain. *Source*: Rodrigue (2020)

sudden change in the availability of raw materials, parts, and manufacturing capabilities. Similarly, demand shocks can be said to be the unavailability of consumer goods (cars, clothing, furniture, or appliances) triggered by a temporary surge in demand toward them. Meanwhile, containment measures such as restrictions on trade, the lack of a workforce (lockdown of workforce), or the closing of key distribution facilities such as airports, ports, and distribution centers would lead the distribution constraints (see Fig. 5).

Global supply chain disruption since the outbreak of COVID-19 was observed more saliently in shipping and port industry than other sectors in international logistics section at least in the first half of 2020, mainly because of the containment measures against the pandemic.

The impacts on the shipping industry can be divided into short- and long-categories. The short-term impacts include manufacturing delays, increased blank sailings, decreasing costs of hiring large crude tankers and the large bulkers (VLCC), slower port operations, occurring cancellations and postponements of cargo transportation, and causing low demand for oil. The long-term impacts are lower contract rates, revenue losses (of liners).

The short-term impacts, in other words, the impacts of initial days of the pandemic outbreak on shipping and port industry were visibly recorded in a Norwegian case study using satellite data tracked by Heiland and Ulltveit-Moe (2020).[10] According to the authors, Norway experienced a significant decline in sailing for all cargo ships (containerships, cargo ships, and cruise ships) mainly due to rapid spread of lock downs—restrictions on vessels that had port calls in the pandemic high-risk countries, and crew who embarked from COVID-19 high-risk countries and crew changes and shore leave (see Fig. 6).

In the first half of 2020, global players in express/parcel industry and land transport sector experienced hardships. On March 19, 2020, the DHL declared the Force Majeure, which a provision in service contracts that under an exceptional condition, service provider can be relieved of liability for non-performance (SupplychainDive, 2020a).[11] In March in the same year, the EU countries' freight forwarders suffered from hard hit of trucking service shortage because of lack of drivers (for drivers' health checks etc.).[12] Most of global freight forwarders such as Agility imposed self-restrictions that if a certain location or site classified by "red zone," the firm automatically cancel business trips to the concerned location, organizing so called "Rapid Response Team" (RRT).[13]

In the second half of 2020 (especially in the 4th quarter of the year), different story, if not completely, has emerged.

In global shipping industry, there are clear signs of freight charge recovery and even shortage of cargo spaces. The signs of recovery began to be observed even in the 3rd quarter of the year. In its technical note published in September of the year, UNESCAP expect a moderate recovery with the reason that the impact of lower demand on freight rates has already been moderated by strict ship capacity management by carriers (UNESCAP, 2020). Blank sailings dominant in global shipping industry in the first half of the year turned to shortage of cargo spaces for a surge of cargo demand.

---

[10]See Baldwin and Evenett (2020), particularly Chapter 11, pp. 151–163.

[11]See the article in SupplychainDive (https://www.supplychaindive.com/news/dhl-global-forwarding-force-majeure-services-covid/574565/).

[12]Consulted from KMI's on-line news letter *Maritime&Fishery foreign information portal*, reflecting monitoring various sources including Ti article (https://www.ti-insight.com/briefs/fred-smith-on-fedex-response-and-experience-with-covid-19/).

[13]The authors' monitoring using official websites conducted in March/April, 2020.

	Short term changes	Potential Long term/Structural change
Shipping & logistics	- Decreased maritime trade - Vulnerable in ports with low trade volume - Due to fleet adjustment including blank sailing - Delayed customs and port clearance - Increased ship waiting time and delaying crew shift - Relatively high freight rates that could further increase if capacity continues to be constrained and demand/supply mismatch occurs	- Facilitated digitalization - Increased risk management costs - Intensified merger and acquisitions and competition among companies - Increased demand for digital professionals - supply chain reconfiguration and design
Ports	- Increase workloads due to the quarantine and health controls - Increased port volume volatility - Shortage of container stacking yard during congestion - Slowdown of port economy - Challenges to smooth inland transport operations and hinterland connections - Changes in port calls and port connectivity - Challenges to yard operations due to peak port calls (i.e. less frequent calls but large port call exchanges) - Storage and warehousing capacity constraints	- Greater use of technology and digital tools to build smart port and automated ports - More focus on increasing resilience and robustness of operation to wide-ranging risks, including from pandemics - Higher focus on port safety and security - Greater use of integrated transport services and inland transport operations - Supply chain reconfiguration and design that may change port network configuration

Fig. 6 Changes in shipping/port sector due to COVID-19. *Source*: UNESCAP (2020)

The similar phenomenon is observed in global port industry. With recovery of demand, container freight rates from Asia to other region began to rise or even surge in the end of 2020, while several ports along the US west coast began to suffer from pressure from enormous leap in cargo throughputs such as LA and Long Beach (The Loadstar, 2020). Cargo traffic of Busan Port, the largest container port in South Korea that went down to by −6.9% by July turned to rise up by +2.2% in October the same year (YONHAPNEWS, 2020).

Newly surged e-commerce related cargo shipments driven by changes of consumers' behavior and consumption pattern (so called "untact economy") led global service providers such as DHL, FedEx, and UPS raise up the expectation of a new boom, increasing their delivery capacities (including aircrafts) and labor force

(SupplychainDive, 2020b). Even in August 2020, Alibaba's online orders began to soar as consumers' shift of shopping during lockdowns (WSJ, 2020).

The short-term scenarios and long-term scenarios are being rewritten and updated, recently, in which one major trait of the change is a newly revised prediction divided into downside and upside, replacing pessimism-oriented prospects dominant in previous predictions, and pointing out rising of some future course possible in the international logistics industry (UNESCAP, 2020) (see Fig. 6).

Despite varying degree of debate on the long-term image of global logistics (such as debate on reshoring issue), there are some consensus among logistics experts that future change would be driven mainly by two commonly shared factors—digitalization and resilience.

3 A New Approach to Supply Chain Risk Mitigation in Era of Pandemics

3.1 Lessons from the COVID-19 Crisis

Before the COVID-19, discussions on mitigation strategy of supply chain risks tended to start a distinction between supply risk and demand risk (Christopher, 2011; Sodhi & Tang, 2012). Based on this distinction, Sodhi and Tang suggest the "Robust" supply chain strategies, which consist of 11 contents including postponement, strategy stock, flexible supply base, etc. (Sodhi & Tang, 2012) (see Fig. 7). But the approach to mitigation strategy didn't work well in the time of the COVID-19.

First, as shown above, disruption occurred by combination of demand shocks and supply shocks in the COVID-19 crisis.

Second, disruptions and following delays in supply chain were rapidly or in an instant, by various risk drivers as seen above—to name a few, human factor that seems to be the labor strikes (shutdown and restriction of labor force's movement); export/import restrictions (as shown in the early stage of the outbreak in some countries movement to food embargo); and conflict and political unrest, which would likely to be escalated at domestic level as well as external level (as for example of SC risk drivers, see Fig. 2 again).

Third, before the COVID-19 crisis, priority of SC risk management aimed at the efficiency represented by minimum costs and minimum lead-time. But after the pandemic, such a priority didn't work under waves of real-time shocks (demand-side and supply-side).

	Robust SC Strategy	Major objective	Benefit(s) under normal risk	Benefit(s) under abnormal risk
1	postponement	Increased product flexibility	supply	Change the configurations of different products quickly
2	Strategic stock	Increases product availability	supply	Respond to market demand quickly during a major disruption
3	Flexible supply base	Increased supply flexibility	supply	Shift production among suppliers promptly
4	Make-and-buy	Increases supply flexibility	supply	Shift production between in-house production facility and suppliers rapidly
5	Economic supply incentives	Increases product availability	supply	Adjust order quantities quickly
6	Flexible transportation	Increases flexibility In transportation	supply	Change the mode of transportation rapidly
7	Revenue management	Increases control of production	demand	Influence the customer Product selection dynamically
8	Dynamic assortment planning	Increases control of production	demand	Influence the demands of different products quickly
9	Silent product rollover	Increases control of Product exposure to customers	Supply and demand	Manage the demands of Different products swifly
10	Flexible supply contracts	Increase Replenishment flexibility	supply	Shift order quantities Across time
11	Flexible manufacturing process	Increase flexibility In producing Different product	demand	Shift production quantities Across internal resources (plants or machines)

Fig. 7 Robust supply chain strategies. *Source*: Sodhi and Tang (2012)

3.2 Challenges for SC Risk Mitigation

3.2.1 Resilience and Agility Coming Back

One of the most commonly discussed words related with SC risk mitigation since the outbreak of the COVID-19 is the concept of the "resilience," which has initially risen in SC related literature in early 2000, and which has prevailed among researchers, business consultants, and industrial commentators. Ivanov (2020) proposes the "viable supply chain model" integrated of agility, resilience, and sustainability. Sneader and Singhal (2020) suggests agility and resilience for success in the "next normal" (later called "new normal"). Guillot (2020) discusses the resilience as an objective achieved by making digital supply chains.

The concept "resilience" in the SC field has been analytically distinguished by Gligor et al. (2019) by comparing it with the concept "agility," and with a thorough literature review. On the one hand, one of key themes of the resilience is the "ability

Agility	Resilience
Ability to quickly change direction	Ability to resist/survive disruptions
Ability to empower the customer/customize	Ability to avoid the shock altogether
Ability to integrate processes within and across firm	Ability to recover/return to original form flowing disruption
Ability to speed/accelerate operations	**Ability to adjust tactics and operations (flexibility)**
Ability to adjust tactics and operation (flexibility)	**Ability to speed/accelerate operations**
Ability to scan the environment/anticipate	**Ability to scan the environment/anticipate**

Fig. 8 Agility and resilience. *Source*: Gligor et al. (2019)

to resist/survive disruptions, while the concept of agility is used to mean mainly "the ability to quickly change direction emerged" as the first theme within agility research (Gligor et al., 2019) (see Fig. 8). On the other hand, the resilience and the agility share a few common characteristics—ability to speed/accelerate operations, ability to adjust tactics and operation (flexibility), and ability to scan the environment/anticipate. It is from the comparison of the concepts that the resilience and agility have a complementary relationship for considering SC mitigation strategies.

3.2.2 Existing SCR Mitigation Strategies and Limitation

When the COVID-19 began its world-wide transmission shifting its nature from epidemic to pandemic, McKinsey researchers suggested a brief package for company's risk practice, of which main points include: protect your employees; set up a cross-functional COVID-19 team; ensure that liquidity is sufficient to weather the storm; stabilize the supply chain (Craven et al., 2020). The authors' recommendations provide an important clue to explore future strategy of SCR caused by the pandemic (not just available to the COVID-19, but also other pandemic in general). When the recommendations started with "protect your employees," it shows that under the outbreak of the pandemic, the human factor itself can be transformed as a potential risk factor (or "risk driver") causing SC disruption and, at the same time, a potential tool for strengthening resilience. It should be the most newly discovered fact for future strategy for coping with the pandemics.

Meanwhile, many SC experts have developed and proposed useful tools for SCR migration. One of them is Jutter et al. (2003), consisted of 4 strategies (avoidance, control, cooperation, and flexibility, henceforth, "ACCF model"), which was modified Miller's "five generic strategies" for company's risk mitigation in SC contexts (see Fig. 9).

For more than two decades, the SC mitigation strategies have been applied to various fields in the international logistics and SCRM, sometimes under the guidance of existing experts' recommendations that mentioned above, sometimes adjusting newly rising external environments with some variation. For example,

Avoidance	- Dropping specific products/geographical markets/supplier and/or Customer organizations
Control	- Vertical integration - Increased stockpiling and the use of buffer inventory - Maintaining excess capacity in production, storage, handling and/or transport - Imposing contractual obligation on suppliers
Cooperation	- Joint efforts improve supply chain visibility and understanding - Joint efforts to share risk-related information - Joint efforts to prepare supply chain continuity plans
Flexibility	- Postponement - Multiple sourcing - Localized sourcing

Fig. 9 Risk mitigating strategies in supply chains. *Source*: Jutter et al. (2003)

the "control" strategy ("maintaining excess capacity in production, storage, handling and/or strategy") was implemented in the form of recommendation for maintaining suppliers (tier-1, -2, -3) and inventory levels, rationing critical parts, prebooking rail/air-freight capacity etc. (Craven et al., 2020, p. 4).

Giannakis and Papadopoulos (2016) is another example as adaptation of the ACCF model, in which the authors, for achieving sustainability, modifies the ACCF into "ACSR" (Avoid, Control, Share, and Retain) as "risk treatment"[14] tools (Giannakis & Papadopoulos, 2016, p. 458). Interestingly, the authors categorize the pandemic as "social-exogenous" sustainable risks with demographic challenges and social instability/unrest (pp. 465–468).

The research examples implicate necessity of rebuilding or reorganize new frameworks more focused on the issues of SCR mitigation (or adaption) of pandemic(s).

3.2.3 Key Words for Considering a New Framework

Preparedness is said to be the best way to reduce pandemic impacts on SC. It is "periodic contingency planning and simulation exercises, covering all relevant hazards and engaging all key sectors." (Jonas, 2012). In logistics and SC field, the preparedness includes pre-allocation of resources and emergency distribution planning as the most visible topics (Queiroz et al., 2020). Like Jonas, the authors of this research put an emphasis on simulations for understanding status quo in SC responsiveness and resilience and prediction of massive disruption scenarios and required performance capacities.

Resilience is one of the frequently focused themes in SC risk disruption and preparedness, mitigation etc. As seen above (Sect. 3.2.1), the resilience can be

[14]In the article's context, the treatment seems to be "mitigation."

translated into the "ability to resist/survive disruption" or "ability to avoid the shock altogether" in SC literature. Jonas as a pandemic expert consider "systemic resilience" as an important tool for "ex post mitigation" measure. In an article published in Harvard Business Review, Willy Shih recommends that for companies to meet the challenges from the COVID-19, it needs to make their supply chain more "resilient" without weakening their competitiveness (Shih, 2020). Queiroz, Ivanov, and Donggui, from their literature review, recommend that the resilience could be categorized—if adapted to the pandemic SC mitigation—into fours: systems, process, control, and recovery (Queiroz et al., 2020). Mckinsey paper grants a new concept to the "resilience" that multi-shoring and reduction of risk by avoiding being dependent on any single source of supply (Sneader & Singhal, 2020).

Digitalization is another important component of new frameworks for building of SCRM in pandemic era. The importance of digitalization and/or the 4th industrial revolution has even repeatedly emphasized in all industrial sector even before the COVID-19. What is the most significant difference between present and past is, if borrowing one commentator Guillot, that "while previously looked to technology such as data, IoT, robotics and analytics to drive efficiency, they're now viewing them as means to increase resiliency in disruptive times." (Guillot, 2020).

This change of idea on and appetite for new technology is due to growing interest in visibility, traceability (Queiroz et al., 2020), while reducing the costs of operations and fostering flexibility (Sneader & Singhal, 2020). Shih grants "new technologies" (automation, new processing technologies, continuous-flow manufacturing, and additive manufacturing) as tools for improving flexibility and lowing costs.

As the so-called "un-tact" consumption and business goes on toward "new normal", digitalization (or digital transformation) can provide international logistics and SCM related service industry with renewed opportunities of growth as already seen in e-commerce proliferation.

Supply chain reconfiguration is a concept related with the vulnerability originated from globalization what is called "the weakness" or "hidden risks." As seen above, the COVID-19 crisis shows that the extended production network for reducing costs (including removing inventory) as well as for accessing markets has vulnerability to sudden SC disruption. As a result, many SC experts began to discuss that SC or network reconfiguration would be unavoidable—if not in short-term, but in the longer-term. For Queiroz, Ivanov, and Donggui, this SC reconfiguration is understand as a tool in SC resilience (Queiroz et al., 2020). Whereas many experts discuss "re-shoring" and "near-shoring," Mckinsey's experts recommends "next-shoring" (move to the location with innovation center or localizing supply chains and creating more collaborative relationships) as an option (Sneader & Singhal, 2020). Cordon expects that SC will experience reconfiguration from under globalization to regionalization for various reasons—eliminating single-source dependencies, and establishing a flexible and adaptable chain, making re-emergence of logistics hub at the regional level (Cordon, 2020).

4 Implications for International Logistics

Since the outbreak of COVID-19, Supply Chain disruption caused by pandemic has been a significantly challenging issue to international logistics service providers and to government policy.

For logistics providers. First, it will be requirement to follow-up digitalization in order to have competitiveness as well as resilience. It is mainly because many shippers have already began their adaption for minimizing the disruptive impacts. Focal areas worth attention in digitalization will be better end-to-end visibility, better predictive analytics, and better and smarter automation.

Second, the international logistic service providers should develop capacities of preparedness such as simulation and contingency plan regularly updated. That would have dual sides of objectives. The one is for themselves, and the other is for their customers, shippers. Eventually such efforts will contribute to enhance their competitiveness.

Third, international logistics service providers need to monitor shippers' movements concerning diversification of suppliers and production bases at global level as well as regional level (re-shoring, near-shoring, and next-shoring). In the future, shippers would likely to accelerate their efforts to make "buffers" in their global supply chains. Based on sharing information they would improve common objectives for response capacities of SC disruption mutually.

Governments need to reflect "new normal" in ways of enhancing resilience in all of the industry and trade—for example, institutionalizing the concept of agility and resilience in relevant industrial policies, plans and strategies. Especially at global SC level, government need to set up a SC reconfiguration plan by careful scenario plannings.

Second, governments need to consider SC mitigation strategies such as avoidance, control, cooperation, and flexibility when reviewing foreign direct investment and governmental spending. For example, a government need to calculate, consulting with shippers and logistics service providers, strategic location and facilities critical for maintaining the country's supply chain, and avoiding SC risks.

Third, governments need to develop various policy measures to support companies' and relevant industrial sectors (for example, shipping and port industry) digitalization efforts—especially small and medium enterprises, and provide information for the firms' self-discovery of business opportunity.

References

Baldwin, R., & Evenett, S. (2020). *COVID-19 and trade policy: Why turning inward won't work*. CEPR Press.

Baldwin, R., & Weder di Mauro, B. (2020). *Mitigating the COVID economic crisis: Act fast and do whatever it takes*. CEPR Press.

Blake, P., & Wadhwa, D. (2020). 2020 year in review: The impact of COVID-19 in 12 charts (worldbank.org)

Christopher, M. (2011). *Logistics & supply chain management* (4th ed.). Pearson.

Cordon, C. (2020). A post COVID-19 outlook: The future supply chain, May 2020. Retrieved June 18, 2020, from https://www.imd.org/research-knowledge/articles/A-post-COVID-19-outlook-The-future-of-the-supply-chain/

Craven, M., Liu, L., Mysore, M., & Wilson, M. (2020). COVID-19: Implications for business, March 2020. Retrieved June 25, 2020, from http://www.amcham-egypt.org/bic/pdf/corona1/McKinsey_Co%20-%20COVID-19-Implications%20For%20Business.pdf

Giannakis, M., & Papadopoulos, T. (2016). Supply chain sustainability: A risk management approach. *International Journal of Production Economics, 171*(2016), 455–470.

Gligor, D., Glior, N., Holcomb, M., & Bozkurt, S. (2019). Distinguishing between the concepts of supply chain agility and resilience: A multidisciplinary literature review. *The International Journal of Logistics Management, 30*(2), 2019.

Guillot, C. (2020). Covid-19 spurs a leap to digital supply chains, *SupplychainDive*, June 9, 2020. Retrieved June 25, 2020, from https://www.supplychaindive.com/news/covid-19-digital-transformation-supply-chains/579452/

Heiland, I., & Ulltveit-Moe, K. H. (2020). An unitended crisis in sea transportation due to COVID-19 restrictions. In R. Baldwin & S. J. Evenett (Eds.), *COVID-19 and trade policy: Why turning inward won't work*. CEPR Press.

Ivanov, D. (2020). Viable supply chain model: Integrating agility, resilience, and sustainability perspectives—Lessons from and thinking beyond the COVID-19 pandemic. *Annals of Operation Research*. Retrieved December 13, 2020, from https://link.springer.com/article/10.1007/s10479-020-03640-6

Jonas, A. (2012). *Pandemic risk*. World Bank. Retrieved December 13, 2020, from https://www.worldbank.org/content/dam/Worldbank/document/HDN/Health/WDR14_bp_Pandemic_Risk_Jonas.pdf

Jutter, U., Peck, H., & Christopher, M. (2003). Supply chain risk management: Outlining an agenda for future research. *International Journal of Logistics: Research & Applications, 6*(4), 197–210.

OECD. (2020). *OECD Economic Outlook*. 108, December 2020. Retrieved January 10, 2021, from https://www.oecd-ilibrary.org/docserver/39a88ab1-en.pdf?expires=1610337790&id=id&accname=oid048561&checksum=C9EC545D211411D3B6546F26074F7775

Queiroz, M., Ivanov, D., & Dongui, A. (2020). Impacts of epidemic outbreaks on supply chains: Mapping a research agenda amid the COVID-19 pandemic through a structured literature review. *Annals of Operation Research*, 16 June 2020. Retrieved January 10, 2021, from https://link.springer.com/article/10.1007/s10479-020-03685-7

Rodrigue, J.-P. (2020). *Impacts of pandemics on supply chains*. Retrieved December 13, 2020, from https://transportgeography.org/contents/applications/transportation-pandemics/impacts-pandemics-supply-chains/

Shih, W. (2020). Global supply chain in a post-pandemic world. *Harvard Business Review*, Sept–Oct, 2020. Retrieved December, 2020, from https://hbr.org/2020/09/global-supply-chains-in-a-post-pandemic-world

Sneader, K., & Singhal, S. (2020). From thinking about the next normal to making it work: What to stop, start, and accelerate, *McKinsey&Company*, May 15, 2000. Retrieved June 18, 2020, from https://www.mckinsey.com/featured-insights/leadership/from-thinking-about-the-next-normal-to-making-it-work-what-to-stop-start-and-accelerate

Sodhi, M., & Tang, C. (2012). *Managing supply chain risk*. Springer.

SupplychainDive. (2020a). Global forwarding declares force majeure, reserving the right to modify services. March 20, 2020. Retrieved January 10, 2021, from https://www.supplychaindive.com/news/dhl-global-forwarding-force-majeure-services-covid/574565/

SupplychainDive. (2020b). DHL Express anticipates 50% volume increase for peak. Oct. 28, 2020. Retrieved January 10, 2021, from https://www.supplychaindive.com/news/dhl-express-peak-season-volume-airfreight/587906/

The Loadstar. (2020). Container freight rates from Asia surge new highs-'it's gone bad.' Dec. 11, 2020. Retrieved December 10, 2020, from https://theloadstar.com/container-freight-rates-from-asia-surge-to-new-highs-its-gone-mad/

UNESCAP. (2020). Covid-19 and its impact on shipping and port sector in Asia and the Pacific, Technical note, 30 September 2020. Retrieved December 10, 2020, from https://www.unescap.org/sites/default/d8files/knowledge-products/ShippingPoliyBrief-16Oct2020-FINAL.pdf

Veselko, G., & Bratkovic, T. (2009). Managing risks and threats in global logistics chains. *Pomorstvo, 23*(1), 67–83.

World Bank. (2020). *2020 Year in Review: The impact of COVID-19 in 12 charts.* Retrieved January 10, 2021, from https://blogs.worldbank.org/voices/2020-year-review-impact-covid-19-12-charts

World Economic Forum. (2012). *New models for addressing supply chains and transport risk.* Retrieved April 10, 2020, from http://www3.weforum.org/docs/WEF_SCT_RRN_NewModelsAddressingSupplyChainTransportRisk_IndustryAgenda_2012.pdf

WSJ. (2020). Alibaba's online orders soar during coronavirus, fueling a sales recovery, Aug. 20. 2020. Retrieved January 10, 2021, from https://www.wsj.com/articles/alibaba-online-orders-fuel-sales-recovery-11597924526

WTO. (2020). *Trade shows signs of rebound from COVID-19, recovery still uncertain.* Press/862 press release. 6 October, 2020. Retrieved January 10, 2021, from https://www.wto.org/english/news_e/pres20_e/pr862_e.htm

YONHAPNEWS. (2020). Container traffic increased first time for 6 months (in Korean), October 13, 2020. https://theloadstar.com/container-freight-rates-from-asia-surge-to-new-highs-its-gone-mad/

Part II
Decarbonization and Green Finance

Container Shipping Decarbonization Pathways

Pierre Cariou and Elizabeth Lindstad

Abstract The International Maritime Organization agreed in 2018 on a strategy to support the long-term goal of a reduction in maritime Greenhouse gas (GHG) emissions by at least 50% by 2050. To achieve this objective, short- and long-term objectives were set. First, to reduce carbon intensity measured in CO_2 per ton-mile by at least 40% by 2030, pursuing efforts toward 70% by 2050, compared to 2008; Second to reduce the total annual GHG emissions by 2050 by at least 50% compared to 2008. This paper assesses whether or not these two objectives are going to be met by 2030 and 2050 for the container shipping segment, which in total accounts for about 30% of GHG shipping emissions by merchant vessels. We estimate actual and future carbon intensity using change over time from a linear extrapolation per ship size, and conclude that the goals on carbon intensity reduction will probably be met by 2030 and 2050. Moreover, we use forecasts on future trade developments to predict future container transport work and to conclude that the objective of halving emissions by 2050 is within reach for container shipping only if trades increase less than 1.5% per year.

1 Introduction

Transportation and logistics are often the greatest source of carbon emissions for companies (Graham et al., 2018) and there is an increasing pressure for reporting carbon footprint (Pazirandeh & Jafari, 2013; Tang et al., 2015). This is a major concern for the shipping community that is a large contributor to worldwide emissions with approximatively 3% of worldwide emissions (Faber et al., 2020).

Furthermore, predictions by OECD (2019) are of an increase in shipping GHG emissions by 23% by 2035 in a business-as-usual (BAU) scenario. When accounting for four different decarbonization pathways, predictions could be reduced from 82% to 95% in 2035 compared to the BAU case, if shipping would implement some

P. Cariou (✉) · E. Lindstad
Kedge Business School, Bordeaux, France
e-mail: pierre.cariou@kedgebs.com

B.-W. Ko, D.-W. Song (eds.), *New Maritime Business*, WMU Studies in Maritime Affairs 10, https://doi.org/10.1007/978-3-030-78957-2_5

short- and long-term measures to reduce carbon intensity (in grams of CO_2 per ton-miles). The latest IMO report (Faber et al., 2020) concludes on forecasts of 90 to 130% in shipping emissions by 2050 compared to 2008, using a range of plausible long-term economic and energy scenarios. Furthermore, if the COVID-19 will lower global emissions for 2020 and 2021, emissions over the next decades may be only a few percent lower than these initial projections.

These predictions have led to a series of private and public initiatives. On the private front, major shipping observer[1] organizations have proposed (ICS, 2019) to create a 5 billion USD fund to finance R&D for technologies and fuels toward decarbonization based on 2 USD per ton of fuel. Earlier, eleven shipping banks agreed on the Poseidon Principles signed in June 2019 to establish a framework for assessing and disclosing the climate alignment of ship finance portfolios. In October 2020 (Offshore Energy, 2020), a group of the world's largest energy, agriculture, mining, and commodity trading companies,[2] including Trafigura, Cargill, and Shell, committed to report on CO2 emissions reduction in line with the IMO's decarbonization goals. Trafigura, one of the largest ship charterers, also sets out a proposal for IMO member states to help accelerate progress towards the decarbonization via a levy on carbon intensive marine fuels of $250–300 per tonne of CO2, in order to close the gap between carbon-intensive fuel and low or zero-carbon fuel.

On the public front, on 15 September 2020, the EU Parliament decided to include Greenhouse Gas (GHG) emissions from ships over 5000 GT by 1 January 2022 in the European emission trading system (EU ETS). Out of the collected money, 50% of revenues from the sale of shipping ETS allowances will be used for an Ocean Fund, established for the period 2022–2030, to improve the energy efficiency of ships (EU, 2019, 2020; ECSA-ICS, 2020). The European and IMO aims are to supports the implementation of carbon-free technologies and the establishment of funding, with a need estimated at $1–1.65 trillion by 2050 or US$ 50–70 bn annually for 20 years (UMAS, 2018, 2020).

In April 2018, the member states of the International Maritime Organization (IMO, 2018) agreed on a strategic plan to reduce annual GHG emissions from international shipping, a plan that should be consistent with the Paris Agreement temperature goals. The strategy can be summarized in three main points. First, to reduce the carbon intensity of ships through implementation of further phases of the energy efficiency design index (EEDI) for new ships. Second, to reduce CO_2 emissions per transport work by at least 40% by 2030, pursuing efforts to reducing them by 70% by 2050 (compared with 2008). Third, to peak GHG emissions from international shipping as soon as possible and to reduce annual GHG emissions by at least 50% by 2050 (compared to 2008). The IMO strategy combined short- and long-

[1]ICS, BIMCO, WSC, Intertanko, Intercargo, Interferry, CLIA and IPTA.

[2]Anglo American, ADM, Bunge, Cargill Ocean Transportation, COFCO International, Dow, Equinor, Gunvor Group, Klaveness Combination Carriers, Louis Dreyfus Company, Norden, Occidental, Shell, Torvald Klaveness, and Trafigura.

term objectives, and a revision towards a more ambitious target will likely be considered in the Revised Strategy due for finalization by 2023 (Koesler et al., 2020).

In this paper, we investigate if these IMO goals are likely to be met by 2030 and 2050. In particular, we assess whether or not the target of a minimum reduction of 40% of CO_2 emissions per transport work (carbon intensity) by 2030 and a reduction of 70% by 2050 both compared to 2008 is realistic. Then, we estimate if the 50% reduction target can be met when account for projection on future container trade.

The chapter focuses on container shipping that provides port-to-port transport of containers at a fixed schedule and with a predetermined transit time (Plum et al., 2014). This focus is justified by the importance of container shipping in international trades, with a share in total tone-miles that has grown from 10% in 2000 to 15% in 2018 (Faber et al., 2020). Container shipping accounts for approximatively 30% of merchant vessels GHG emissions in 2012 and 2018 and 20–25% of all GHG shipping emissions estimated at 1 billion tons per year (Faber et al., 2020).

This chapter investigates how carbon intensity and transport work may change up to 2050. In relation to existing literature and to the IMO initial strategy, the main contributions are to:

- Provide an evaluation of carbon intensity based on the Annual Efficiency Ratio (AER) that gives the possibility to understand how much improvements have occurred from 2008 to 2018 and what are the underlying factors (technology, operational improvement and economies of scale) behind these developments.
- Provide an evaluation of total carbon emissions in 2018 and in 2050, by containership sub-segment, and to predict whether or not the global IMO objective of halving container shipping emissions by 2050 is realistic.

The chapter is organized as follows. Section 2 presents the main factors that explain container shipping emissions. Then, Sect. 3 presents our method to measure changes in carbon intensity up to 2018, and an evaluation on whether or not the decrease in carbon intensity goals up to 2030 and 2050 is realistic. Section 4 uses prospects on seaborne trade development to estimate future container shipping emissions by 2050 and to evaluate whether or not the global objective of a 50% reduction goal is achievable. Section 5 is our conclusions on the implementation of the IMO initial strategy.

2 Literature Review on Containerships Decarbonization

An extensive literature exists on long term shipping decarbonization pathways (OECD, 2019; Buhaug et al., 2009; Smith et al., 2014; Raucci et al., 2017; UMAS, 2018, 2020) and on CO_2 shipping emissions (Corbett et al., 2009; Lindstad et al., 2011, 2013, 2015a, 2015b; Cariou, 2011; Cariou et al., 2019; Olmer et al., 2017; Faber et al., 2009, 2012, 2017; CSC, 2017; Comer et al., 2018; Leaper, 2019; Wang & Meng, 2012; Psaraftis & Kontovas, 2013, 2014; Psaraftis, 2019; Faber et al., 2020). In short, changes in global emissions depend on the evolutions in carbon intensity (in grams of CO_2 per ton nautical miles) and in transport work (in ton nautical miles per year).

Improvements in carbon intensity can be achieved through technological, operational changes and economies of scale (Psaraftis, 2016; Bouman et al., 2017; OECD, 2019; Balcombe et al., 2019). To achieve large improvements such as the IMO target of 50 percent GHG reductio, a switch to low or zero carbon fuels is then needed (Hoegh-Guldberg et al., 2019). In container shipping, Cariou et al. (2019) identify six factors driving changes in CO_2 emissions (Fuel Type & Efficiency, Vessel Size; Vessel Number; Number of Port & distance travelled; Commercial Speed and Time in Port). We grouped these six generic factors into 3 main categories.

First, technological-related improvements (Fuel Type & Efficiency) which are related to propulsion, machinery systems and hull design. In addition, modern hull shapes of mega-ships are optimized for reduced drag, thus lowering fuel consumption and CO_2 emissions (Stott & Wright, 2011; CCNR, 2012; Gilbert et al., 2014). The utilization of cleaner fuel types to meet the regulatory stipulations of IMO (e.g. Marpol Annex VI) coupled with more efficient engines have also the potential to reduce emissions considerably (Wärtsila, 2009; Wang & Lutsey, 2013; Tillig et al., 2015). The switch to zero or lower carbon content fuels can be done through different options[3] (Hanson et al., 2020; Balcombe et al., 2019; DNV, 2019) and out of seven zero-emission technology evaluated (UMAS, 2018, 2020) the E-fuels, i.e. liquid or gaseous fuels made from renewable electricity are favored over the Biofuels.

Second, carbon intensity improvements related to economies of scale, that could be deployed thanks to network-related measures which consist in playing with the number of port of calls, the distance to travel, time in port, which give the possibility to deploy larger vessels in order to take advantage of economies of scale on fuel consumption (Imai et al., 2006; Tran & Haasis, 2015). These changes require some improvements in port productivity (Time in port) that can be achieved, for instance, by progress in digitalization, by the use of bigger and faster ship-to-shore cranes, a higher crane density per ship, or lower dwell times (Felício et al., 2015; Ha et al., 2017).

Third, additional reduction in carbon intensity can be achieved through operational improvements, with the optimization and with the decrease, for instance, of the average commercial speed of vessels (Corbett et al., 2009; Lindstad et al., 2011; Wang & Meng, 2012; Meng et al., 2016; Wang et al., 2018; Adland et al., 2018, 2020; Du et al., 2019). The introduction of slow steaming practices dates back to 2008 when shipping lines needed to reduce bunker consumption to drastically cut operating costs and it had a significant impact on the reduction in carbon emissions

[3]Namely electric, hybrid hydrogen, hydrogen fuel cell, hydrogen +ICE, ammonia, ammonia + ICE and biofuel.

from 2008 to 2012 (Notteboom & Vernimmen, 2009; Cariou, 2011; Psaraftis & Kontovas, 2013; Bouman et al., 2017).

Transport-work is the second main determinant to understand changes in future global emissions. Transport-work is obviously related to changes in seaborne demand (Smith et al., 2014; DNV, 2017; UNCTAD, 2017, 2018a, 2018b; OECD, 2018; ITF, 2019; CE Delft, 2019; Cariou, 2019; Faber et al., 2020). According to Smith et al. (2014), due to the increase in trade, emissions in 2050 could be similar to 2008 and in the worst case, to three and half time the 2008 value (Buhaug et al., 2009; Eyring et al., 2010). Lindstad et al. (2018) predict a doubling of emissions by 2050 compared to 2010, when assuming a 3% yearly increase in transport work and 1% energy efficiency improvement. Traut et al. (2018) rely on Smith et al. (2014) projections on future sea transport demand to conclude a 8.7% reduction per annum in global fleet carbon intensity is needed to reach Paris agreement goals, and 2.1% for dry bulk carriers and 2.2% for wet bulk carriers, rates, which are higher than former estimates of 2.1 to 3.3% reduction per annum for the World fleet (Buhaug et al., 2009) or 2.5% p.a. (Smith et al., 2014).

In the specific container segment, forecast on future demand is high according to DNV (2017), compared to other segments. It follows the main trend in GDP growth (Cariou, 2019). and Fontagné and Fouré (2013) study on the long-term perspectives of the World trade stresses a structural change that has occurred during last decades. Since 1950, trade used to increase at a faster path than industrial or agricultural production, and even more than GDP, illustrated by a long-term elasticity with respect to GDP at 1.64 over the period 1950–2009. The latest period shows a reduction from 2.82 in 1990–1999 to 1.42 in 2000–2009. This element is also underlined by UNCTAD (2017) that mentions a trade-GDP elasticity at 1.3 in 1970–1985, 2.2 in 1986–2000, 1.3 in the 2000s and 0.7 in 2008–2013.

According to DNV (2017), the container trade will experience, as measured by tone-miles, an average 3.2% increase per year up to 2030. It will thereafter decline to an average 2.1% per year. Over the entire forecasting period to 2050, the annual growth will average 2.6% for container tone-miles and 2.4% for global GDP or a container trade-GDP multiplier (trade growth relative to GDP growth) at 1.1. Container growth is projected for all regions, and the Indian Subcontinent is subject to the strongest growth, followed by China, South East Asia and Sub-Saharan Africa.

Many other drivers will also impact future container trades (Cariou, 2019). The ITF Transport outlook (2019) concludes that under current situation freight growth will triple by 2050, and that maritime transportation will continue to dominate freight (70% of total freight with 120 000 billion ton-km by 2030 and 250 000 by 2050). The growth will be mostly supported by Asia, Africa, Indian and pacific oceans. However, current and future disruptions related to logistics efficiency and technology (e-commerce, 3D printing, new trade routes, energy transition) could lower maritime demand and the increase in total maritime demand from 2019 to 2050 would be at approximatively 2.3% per year compared to 3.8% for the last 50 years. For DNV (2017), particular attention will be given to decarbonization and to the development of the circular economy that could represent a major game changer. Future trade volume, pattern and modal shifts could also be significantly impacted by the development of new routes, such as the One Belt One Road for rail or by the opening of Arctic shipping.

Technology development are another important factor. McKinsey (2019) stresses for instance that the adoption of electric vehicles will disrupt automotive value chains and trade as battery-powered electric vehicles have only 20 to 30 moving parts in their drivetrains, compared to 130 to 170 moving parts in an internal combustion engine. Additive manufacturing and 3D printing will reduce the need for long-distance trades as on-demand production near the consumer replaces the global distribution of mass-produced goods. The increased use of robots could enable relocation of production back to developed countries, shortening global value chains, and potentially reducing demand for seaborne transport.

The Danish Ship Finance (2018) analysis on future maritime outlooks predict a long-term growth in seaborne trade volumes averaging about 1% per annum until 2030 which is about half of DNV forecasts for 2015–2030. A shift towards different vessel types, smaller parcel sizes and in some segments even fewer cargoes shipped is expected, due to the combined effects of the fourth industrial revolution (e.g. artificial intelligence, robotics, the internet of things, 3D printing and digitalisation) and the ageing consumer base. These changes are particularly important for containerized shipping markets which heavily rely on good producing GVCs (McKinsey, 2019).

3 Past and Future Changes in Container Carbon Intensity

The vessel carbon intensity is the amount of CO_2 generated to transport a given quantity. There is a wide range of indicator of carbon intensity (Panagakos et al., 2019; Pereira, 2019a, 2019b, 2020; Faber et al., 2020) and we rely in this chapter on the Annual Efficiency Ratio (AER) which are defined as follows:

$$AER = \frac{\sum_{j} FC_j.EF_j}{DWT.Dist} \tag{1}$$

where FC_j is the mass of fuel type j consumed, EF_j the emission factor for fuel j, DWT is the total Deadweight tons carrying capacity and $Dist$ the distance sailed in nautical miles. Compared to other indicator such as the Energy Efficiency Operational Indicator (EEOI), AER has the advantage since there is no need to convert volume-based cargoes (TEU for instance) to tons, meaning for container shipping no need to estimate the number of empty containers.

Two main sources of information were used to estimate AER. For vessels operating in 2018, the data are based upon performance as reported per ship type and size as reported to the EU Thetis-MRV data collection system for 2018 (Pereira, 2019a, 2019b, 2020; European Commission, 2020). The data for 2008 and 2102 data are based on the Second IMO GHG study (Buhaug et al., 2009) and the third IMO GHG study (Smith et al., 2014).

The main difference between the IMO study and the EU Thetis-MRV is that the former estimates emissions per ship segment based on activity level (number of days at sea, in port, speed, number of vessels…) from Automated Information System (AIS) and naval architects' assumptions on fuel consumption modelling while the MRV relies on data that are directly reported by ship operators, certified by independent accreditors (Panagakos et al., 2019). A comparison of fuel consumption level estimated (Faber et al., 2020) concludes to differences in AER between the two method of around −0.2% for bulk carriers and 3% for oil tankers, while error is larger for other vessels types such as for container vessels (6%).

In our work, the AER are estimated from the relationship of AER to DWT through regression curves, with results reported in the next figures for 2008 (using IMO, 2009), for vessels built up to 2012 (IMO, 2014) and for vessels built after 2012 (EU Thetis MRV, 2018). Comparing the two IMO studies, the Second IMO study provided the AER as a function of ship type and main size groups, while the Third study provided neither. However, with the AER it is straight forward to recalculate the AER values simply by dividing the total reported CO_2 emissions per vessel type and size on the reported days at sea and average speed.

Findings for containerships are reported in Fig. 1, where the graph reports the changes in carbon intensity or AER in grams of CO_2 per ton nm on the y-axis with vessel size measured as dwt on the x-axis. It shows estimations on AER in 2008 to 2018 with a split between vessels built up to 2012 on the left side and vessels built after 2012 on the right side. Table 1 identifies the three main determinants that can explain the change in carbon intensity: reduction in operational improvement (mostly through speed reduction); design & technical improvements and economies of scale.

During the last 10 years (2008 to 2018), the average AER has decrease by 27.1% from 2008 to 2018. For vessels built before 2012, the improvement is related to operational (12.9%) and from economies of scale (12%). When considering vessels built after 2012, the global reduction in AER is higher than average and estimated at 30.7%. Economies and scale (12.5%) and operational improvements (+12.9%) such as those related to the decrease in operational speed are the first two determinants, while design-technical improvement have a more limited effect (+12.5%).

These estimates stress that the global objective of at least 40% reduction in container carbon intensity by 2030, and by 70% for 2050 are achievable, when relying on changes observed in AER from 2008 to 2018. These findings are in line with previous studies of emissions abatement measures in shipping (Bouman et al., 2017) that have documented that it is possible to improve energy efficiency and reduce emissions in a cost-effective manner, either with zero costs or with net cost savings (Alvik et al., 2010; Buhaug et al., 2009; Faber et al., 2009, 2011). In Bouman et al. (2017) the review on the reduction potential of 22 individual measures grouped under five main categories (*hull design, power and propulsion, alternative fuels, alternative energy sources* and *operations)* leads the authors to conclude that, in total, there is a potential for reducing emissions by 75–85%, i.e. a factor of 4 to 6 per

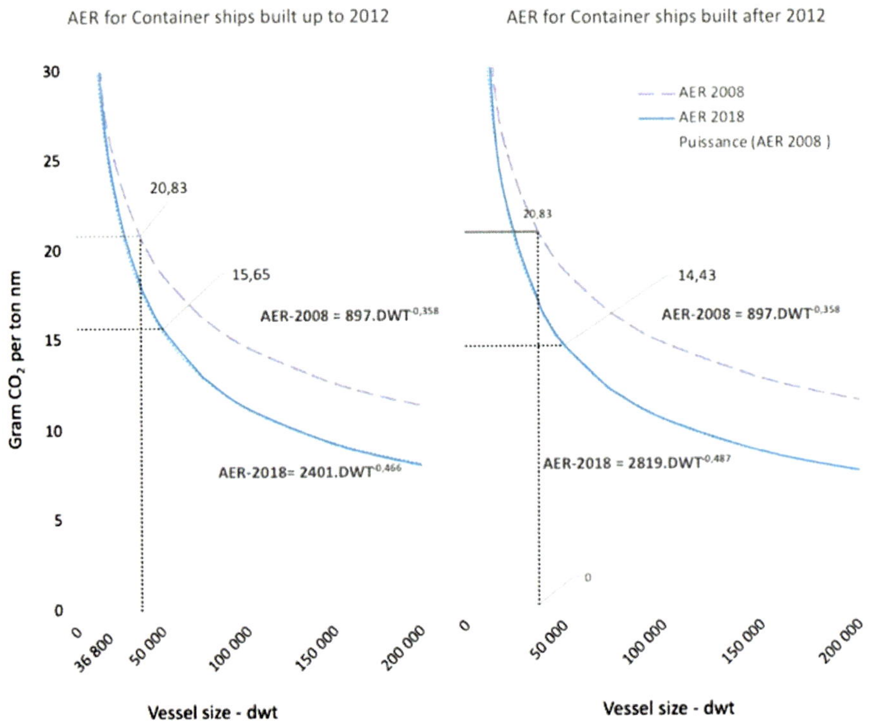

Fig. 1 Development of AER 2008–2018 for containerships. *Source:* Authors

Table 1 Determinants of changes in containerships carbon intensity (AER)

All vessels in 2018		
Average size (dwt) 2008	36 800	
Average size (dwt) 2018		50 000
AER 2008 (gram CO2/ton nm)	20.83	
AER 2018 average (gram CO2/ton nm)	17.62	15.19
Total reduction		27.1%
AER 2018 for vessels built up to 2012	18.15	15.65
AER 2018 for vessels built after 2012	17.24	14.43
Vessels built up to 2012		
Operational improvement	12.9%	
Total reduction		24.9%
Vessels built after 2012		
Operational improvement		12.9%
Design-technical improvement		5.4%
Economies of scale		12.5%
Total reduction		30.7%

Source: Authors' calculations

freight unit transported, based on current technologies and based on the 3rd quartile values. When using the median values, an emission reduction of between 50 and 60% per freight unit transported up to 2050 is more realistic.

4 Past and Future Container Global CO_2 Emissions

To estimate total carbon emission per ship segment in 2008 and 2018, we use our former regression line for AER in 2008 for vessels built after 2012. The average size per segment, is similar to Smith et al. (2014) and Faber et al. (2020) and is applied to eight different container categories (in TEU) and in using the average DWT per ship category. The conversion of our estimates into annual CO_2 emissions accounts for the transport work (Nb vessels × DWT × average Speed × Days at sea) per ship segment. For 2008, we used Smith et al. (2014) and for 2018, figures reported in Faber et al. (2020). When comparing our estimates in 2008 and 2018 with those reported by IMO, there is a difference of 0.8% in 2008 and −3% in 2018 in total emissions, equivalent to 1.6 M. tonnes of CO_2 in 2008 (out of 211 Million) and 6.3 M tonnes of CO_2 in 2018 (out of 235 Million) (Table 2).

For future transport work up to 2050, we considered the main drivers of container shipping markets (UNCTAD, 2018a, 2018b; Cariou, 2019). Future prospects mostly depend on whether or not the GDP per capita of developing countries will increase at the same rate than in former years.[4] Leimbach et al. (2017) analysis on future GDP per capita for the five Shared Socio-economic Pathways SSP1-SSP5 (IPPC, 2000; van Vuuren et al., 2014, 2017) conclude on an increase from 3% (SSP1) to 1.9% (SSP2) for 2010 to 2040, and from 2.2% (SSP1) to 0.5% (SSP3) for 2040 to 2050. Most long-term forecasts agree on a convergence over time amongst higher, middle and lower countries (Johansson et al., 2012) with a growth of non-OECD countries outpacing OECD countries, but with a difference narrowing over the coming decades (from 7% per year over the last decade, non-OECD growth will decline to around 5% in the 2020s and to about 2.5% by the 2050s).

Fontagné and Fouré (2013) on the long-term perspectives for World trade also stress a structural change with a decrease in long-term elasticity with respect to GDP at 1.64 over the period 1950–2009, 2.82 in 1990–1999 and 1.42 in 2000–2009. This element is also found by UNCTAD (2017) that estimates a trade-GDP elasticity at 1.3 in 1970–1985, 2.2 in 1986–2000, 1.3 in the 2000s and 0.7 in 2008–2013. To base our forecast of transport-work up to 2030, we rely on predictions from DNV-GL (2017) and IMO (Faber et al., 2020) that are both based on historical data and hypothesis on socio-economic scenarios.

There is a large range of outcomes according to the method used and hypothesis on future economic development. Faber et al. (2020) report various logistics

[4]Without obviously mentioning the unexpected short-term effects imply by external shocks such as the COVID-19.

Table 2 Comparison of containerships global emissions in 2008 and 2018

	2008		2018	
Size in TEU	IMO	Our estimates	IMO	Our estimates
0–999	14 564 000	14 988 203	10 200 000	9 759 190
1′–1999	32 098 000	31 743 738	28 500 000	27 276 553
2′–2999	29 362 000	28 875 221	21 200 000	22 106 788
3′–4999	60 102 000	61 526 434	40 100 000	38 922 918
5′–7999	53 345 000	49 896 085	41 300 000	36 382 844
8′–11999	22 853 000	23 647 101	58 800 000	52 575 986
12–14500	1 237 000	1 280 371	22 300 000	21 210 436
14500+			13 400 000	21 210 436
Total	213 561 000	211 957 153	235 800 000	229 445 153

Source: Authors' calculations

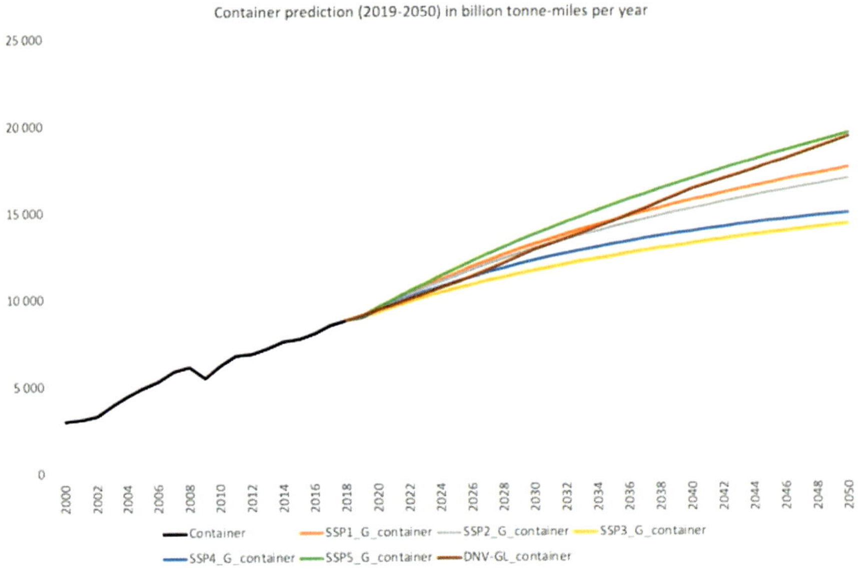

Fig. 2 Future transport work container prediction (1983–2050). *Source:* Authors calculations based on Faber et al. (2020), Cariou (2019) and DNV (2017)

predictions and gravity model projections. Figure 2 presents past evolution and predictions for the RCP4.5 scenario (medium-low mitigation measure for carbon) for the five SSP, restricting our attention to estimates from the Gravity model.

Annual growth rate in transport work (Table 3) ranges from 1.5 to 2.5% per year from 2019 to 2050. We then built two scenarios corresponding to these two situations to estimate the potential changes in global containership emissions up to 2050, and when assuming that in line with former results, the carbon intensity per ship-segment will be at 70% of 2008 value in 2050 (Sect. 3). Figure 3 presents the result for our two scenarios and Table 4 shows the predictions per ship segment.

Table 3 Average annual growth rate of container trade predictions (2019–2015)

	Faber et al. (2017)					OECD (2019)	DNV (2017)
	SSP1_G	SSP2_G	SSP3_G	SSP4_G	SSP5_G		
AAGR	2.2%	2.0%	1.5%	1.6%	2.5%	2.0%	2.5%

Source: Authors calculation from Faber et al. (2020), Cariou (2019) and DNV (2017)

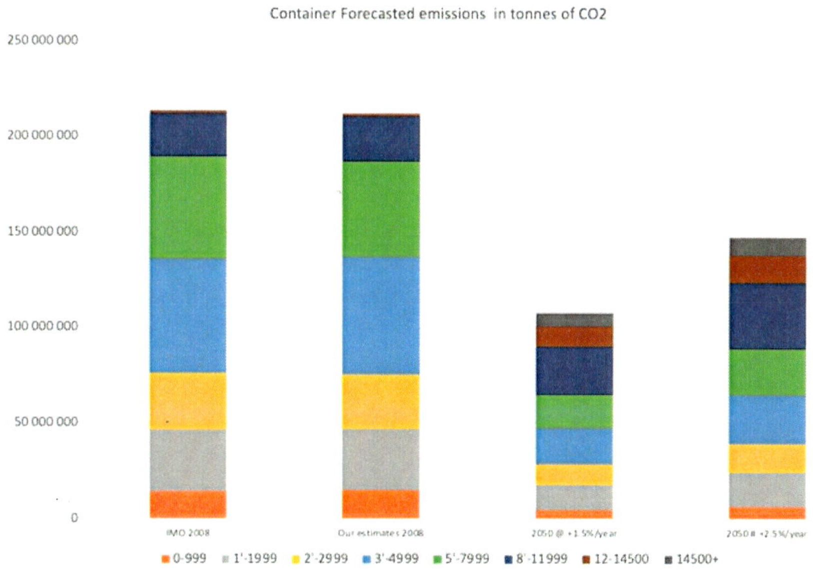

Fig. 3 Containerships CO_2 emissions projections for 2050. *Source:* Authors calculation

The main observations from Fig. 3 and Table 4 are: First, when assuming a 70% improvement in carbon intensity by 2050, the goal of 50% reduction in total emissions by 2050 would be reached (−49.2%) when we assume an annual average growth rate of 1.5% from 2019 to 2050.

When assuming an AAGR of 2.5% per year over the period, the decrease in total emissions is about 30.5%. As reported in Table 4, most of the decrease is explained by changes that occur for vessels less than 8000 TEU, while for larger unit and due to the increase in transport work for these large containerships and over time, there is a significant increase in emissions from 77% to 144% for scenario 1 and 2. When comparing our findings with Bouman et al. (2017) review on annual fleet-wide CO_2 emission estimates available in fourteen studies (Table 5 and Fig. 4) which are also based on adoption of technical and operational measures and on potential growth in demand, our results are similar to Lindstad et al. (2015b), Eide et al. (2011) or CCNR (2012), although the scope and the initial and final year under study varies from one study to another.

According to the latest IMO study (Faber et al., 2020), the general objective of a reduction of 50% of global emission is also achievable, as long as, about 64% of the

Table 4 Future containerships global emissions in 2050

	2008	2050 @ +1.5%/year	2050 # +2.5%/year
0–999	14 988 203	4 714 638	6 452 065
1′–1999	31 743 738	13 177 229	18 033 268
2′–2999	28 875 221	10 679 729	14 615 396
3′–4999	61 526 434	18 803 556	25 732 995
5′–7999	49 896 085	17 576 454	24 053 684
8′–11999	23 647 101	25 399 317	34 759 408
12–14500	1 280 371	10 246 705	14 022 794
14500+		7 021 887	9 609 575
Total	211 957 153	107 619 516	147 279 186

Source: Authors calculation from Faber et al. (2020), Cariou (2019) and DNV (2017)

total amount of CO_2 reduction is contributed to by use of alternative fuels. Under this assumption, and when considering the abatement potential and costs of 44 technologies in four group (energy-saving technologies; use of renewable energy; use of alternative fuels; and speed reduction) and applying all the potential mitigation measures selected to all newly built ships from 2025, CO_2 emissions reduction in 2050 can achieve both the mid-term and long-term levels of ambition specified in the Initial IMO Strategy on Reduction of GHG Emissions from Ships.

5 Conclusions

The objective of this paper was to assess whether or not the main objectives set by the IMO 2018 initial strategy will be met by 2050. First, a reduction of at least 40% reduction in CO_2 per ton-mile by 2030 and 70% by 2050 compared to 2008. Second, the objective of reducing global shipping emissions by 50% by 2050.

Our main conclusion is that for container shipping, the objectives on carbon intensity will probably be reached, while the goal of halving global emission by 2050 will only be met under the assumption of 1.5% annual average increase in container trades (lower forecasts in this study).

Our results remain however highly dependent on assumption on the deployment of new abatement measures to reduce carbon intensity and on future trade development, and are subject to limitations. First, as our evaluation is for CO_2 which does not account for all GHGs. Second, as uncertainties exist on some of our assumptions used to forecast future trades, as in any forecasting exercises, although a consensus seem to emerge in predictions for approximatively 1.5 to 2.5% average growth rate. Third, many operational factors that play on fuel consumption and therefore on emissions, such real conditions at sea, are not considered (Lindstad et al., 2019). Fourth, our evaluation is for container vessels, which account for a significant proportion of shipping-related CO2 emissions (25%) but is not considering the entire World fleet.

Table 5 Potential reduction in CO_2 emissions from shipping at fleet level

Authors	Method	Application	Ref. year	Final year	Mt CO2	Reduction (%)
Buhaug et al. (2009)	Expert estimates, fleet development scenario model 25 measures	18 ship types world fleet	2007	2020 2050	1294 2595	**16–34%** **58–75%**
Faber et al. (2009)	Same as Buhaug et al. (2009) 29 measures	9 ship types, 40 055 ships	2006	2030	1790	**27–47%**
Alvik (2010)	Expert estimates 25 measures	59 ship segments, world fleet	2009	2030	1530	**56%**
Wang et al. (2018)	Similar to Buhaug et al. (2009) 22 measures	14 ship types, world fleet	2007	2020 2030	1421 2205	**35–45%** **35–45%**
Faber et al. (2011)	Updated of Wang et al. (2018)	14 ship types	2007	2020 2030	1204 1982	**20–46%** **20–46%**
Eide et al. (2011)	CO_2 emission scenario model 25 measures	59 ship types, world fleet, 59 800 ships	2008 2008	2020 2030	1191 1533	**41%** **56%**
Hoffmann et al. (2012)	Cost-effectiveness. 25 measures	59 ship segments, world fleet	2010	2030	1316	**30–53%**
CCNR (2012)	Model-based. 27 measures	Inland navigation, Western Europe	2010	2050	2.9	**58%**
Eide et al. (2013a)	Same as Eide et al. (2011) 27 measures	59 ship segments, world fleet, 59 800 ships	2008	2050	2000	**55–95%**
Eide et al. (2013b)	Same as Eide et al. (2011) 15 measures (existing ships)	59 ship segments, world fleet, 59 800 ships	2010	2010	999	**25–32%**
Lindstad (2013)	Model-based, 5 main measures	11 ship types, 109 000 bt nm	2007	2050	2199 2650	**77%** **76%**
Kollamthodi et al. (2013)	Scenario model, 3 main measures	European shipping	2010 2010	2030 2050	279 403	**20%** **33%**
Heitmann and Peterson (2014)	Abatement costs scenarios, 22 measures	14 ship types	2012	2020	947	**22–48%**
Lindstad et al. (2015b)	Model-based, 12 measures (existing ships)	10 ship types, world fleet	2012	2030	2007	**30–45%**

Source: Bouman et al. (2017)

Apart from these methodological aspects, our article sheds light on differences in the future development of CO_2 container emissions per ship segment. It raises the question of the accuracy of having similar objectives and on the need to change targets according to the type of vessels. It therefore means that for container trades which are the largest contributor to emissions, some specific challenges might exist

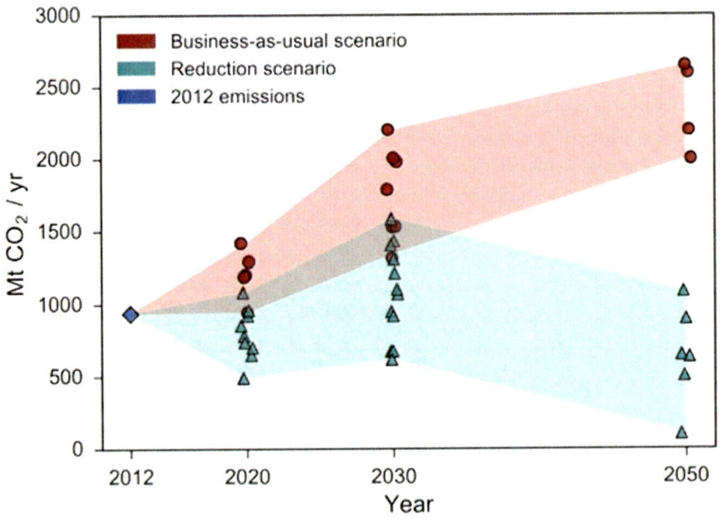

Fig. 4 Annual CO_2 emissions from the global shipping fleet, distinguished by business-as-usual and reduction scenario pathways (Bouman et al., 2017)

in the future in order to be aligned with the IMO objectives. The choice over the type of fuel selected by the main carriers in the next decade is therefore critical. For instance, a major move toward LNG, which is relevant when considering the reduction in non-GHG such as Sox of PM, but which as a limited effect on CO_2 emissions (approximatively less 15–20%) could for instance reduce the likelihood of reaching the long-term objective.

Finally, as shown in this chapter, it appears that reducing emissions for vessel types subject to high demand (containership, cruise vessels, ferries. . .) is obviously more difficult to reach. It confirms that achieving the long-term IMO goal require a massive implementation of zero or near-zero carbon emission technologies for vessels subject to a significant (more than 2%) increase in future trades. In this regard, the change in trade pattern should be as much important as the change in technology to reach the long-term objective. The transition toward more regional trades, due to the change in consumer behavior (time-to-market) and to the rise of protectionism could reduce demand in tons nautical miles and help reaching the 50% reduction goal. This general finding could be the subject of future research on ship-specific goal-based measures to set before 2023 and to account for market dynamic that are currently under discussion at IMO.

Acknowledgment This research has been partly supported by the Social Sciences and Humanities Research Council of Canada (SSHRC) project (N° 895-2017-1003) on "Green Shipping: Governance and Innovation for a Sustainable Maritime Supply Chain" and it was first presented at the Korean Maritime Institute-World Maritime University joint seminar (November 12–13, 2020).

References

Adland, R., Cariou, P., & Wolff, F. C. (2020). Optimal Ship speed and the cubic law revisited: Empirical evidence from an oil tanker fleet. *Transportation Research Part E, 140*: Forthcoming.

Adland, R., Cariou, P., & Wolff, F. V. (2018). The energy efficiency effects of periodic ship hull cleaning. *Journal of Cleaner Production, 178*, 1–13.

Alvik, S., Eide, M. S., Endresen, O., Hoffmann, P., & Longv, T. (2010). *Pathways to low carbon shipping-abatement potential towards 2030*. DNV.

Balcombe, P., Brierley, J., Lewis, C., Skatvedt, L., Speirs, J., Hawkes, A., & Staffell, I. (2019). How to decarbonize international shipping: Options for fuels, technologies and policies. *Energy and Conservation Management, 182*, 72–88.

Bouman, E. A., Lindstad, E., Rialland, A. I., & Stomman, A. H. (2017). State-of-the-art technologies, measures, and potential for reducing GHG emissions from shipping – A review. *Transportation Research Part D, 52*, 408–421.

Buhaug, Ø., Corbett, J. J., Endresen, Ø., Eyring, V., Faber, J., Hanayama, S., Lee, D. S., Lee, D., Lindstad, H., Markowska, A. Z., Mjelde, A., Nelissen, D., Nilsen, J., Pålsson, C., Winebrake, J. J., Wu, W., & Yoshida, K. (2009). *Second IMO GHG Study 2009*. IMO.

Cariou, P. (2011). Is slow steaming a sustainable means of reducing CO2 emissions from container shipping? *Transportation Research Part D, 16*, 260–264.

Cariou, P. (2019). Changing demand for maritime trades, OECD/ITF Round table on the future of Maritime Trade Flows, Paris, France, 15–16 April 2019.

Cariou, P., Parola, F., & Notteboom, T. (2019). Towards low carbon global supply chains: A multi-trade analysis of CO_2 emission reductions in container shipping. *International Journal of Production Economics, 208*, 17–28.

CCNR. (2012). Possibilities for reducing fuel consumption and greenhouse gas emissions from inland navigation. Report by the Inspection Regulations Committee for the 2012 Autumn Meeting. Annex 2 to protocol 2012-II-4 of the Central Commission for the Navigation of the Rhine.

CE Delft. (2019). *Update of maritime greenhouse gas emission projections*. January, CE Delft.

Comer, B., Chen, C., & Rutherford, D. (2018). Relating short-term measures to IMO's minimum 2050 emissions reduction target, in Appendix to paper MEPC 73/INF.27 presented to IMO Marine Environment Protection Committee. 73rd session, (London).

Corbett, J. J., Wang, H., & Winebrake, J. J. (2009). The effectiveness and costs of speed reductions on emissions from international shipping. *Transportation Research Part D, 14*(8), 593–598.

CSC. (2017). Global speed management as an important component of an ambitious initial IMO GHG Strategy, in Paper ISWG-GHG 2/2/16 presented to IMO Working Group on Reduction of GHG Emissions from Ships. 2nd session, (London).

Danish Ship Finance. (2018). Maritime Trend Report: Observations and perspectives on the future of the maritime industry. https://www.shipfinance.dk/media/1910/maritime-trend-report.pdf

DNV-GL. (2017). Maritime Forecast to 2050, Energy Transition Outlook 2017. https://eto.dnvgl.com/2017/maritime

DNV-GL. (2019). *Comparison of alternative marine fuels*. DNV GL AS Maritime.

Du, Y., Meng, Q., Wang, S., & Kuang, H. (2019). Two-phase optimal solutions for ship speed and trim optimization over a voyage using voyage report data. *Transportation Research Part B, 122*, 88–114.

ECSA-ICS. (2020). Implications of application of the EU Emissions Trading System (ETS) to international shipping, and potential benefits of alternative Market-Based Measures (MBMs). Available at: https://www.ics-shipping.org/docs/default-source/resources/environmental-protection/ics-ecsa-study-on-eu-ets.pdf?sfvrsn=10

Eide, M. S., Longva, T., Hoffmann, P., Endresen, Ø., & Dalsøren, S. B. (2011). Future cost scenarios for reduction of ship CO2emissions. *Maritime Policy & Management, 38*(1), 11–37.

Eide, M. S., Dalsøren, S. B., Endresen, Ø., Samset, B., Myhre, G., Fuglestvedt, J., & Berntsen, T. (2013a). Reducing CO2 from shipping – do non-CO2 effects matter. *Atmospheric Chemistry and Physics, 13*, 4183–4201.

Eide, M. S., Chryssakis, C., & Endresen, Ø. (2013b). CO2 abatement potential towards 2050 for shipping, including alternative fuels. *Carbon Management, 4*(3), 275–289.

European Commission. (2020). *2019 annual report on CO2 emissions from maritime transport*. Retrieved at https://ec.europa.eu/clima/news/commissionpublishes-first-annual-eu-report-co2-emissions-maritime-transport_en

EU. (2019). Communication from the Commission to the European Parliament, th European Council, the Council, the European Economic and Social Committee and the Committee of the Regions "the European Green Deal" (Com/2019/640 Final).

EU. (2020). *Annual report on CO2 emissions from maritime transport*. Retrieved at: https://ec.europa.eu/clima/news/commission-publishes-first-annual-eu-report-co2-emissions-maritime-transport_en

EU Thetis MRV. (2018). *CO2 emission report*. Available at https://mrv.emsa.europa.eu/#public/eumrv

Eyring, V., Isaksen, I. S., Berntsen, T., Collins, W. J., Corbett, J. J., Endresen, O., Grainger, R. G., Moldanova, J., Schlager, H., & Stevenson, D. S. (2010). Transport impacts on atmosphere and climate: Shipping. *Atmospheric Environment, 44*, 4735–4771.

Faber, J., Hanayama, S., Zhang, S., Pereda, P., Comer, B., Hauerhof, E., Schim van der Loeff, W., Smith, T., Zhang, Y., Kosaka, H., Adachi, M., Bonello, J.-M., Galbraith, C., Gong, Z., Hirata, K., Hummels, D., Kleijn, A., Lee, D. S., Liu, Y., Lucchesi, A., Mao, X., Muraoka, E., Osipova, L., Qian, H., Rutherford, D., Suárez de la Fuente, S., Yuan, H., Velandia Perico, C., Wu, L., Sun, D., Yoo, D.-H., & Xing, H. (2020). *Fourth IMO GHG Study 2014*. IMO.

Faber, J., Huigen, T., & Nelissen, D. (2017). *Regulating speed: A short- term measure to reduce maritime GHG emissions*. CE Delft publication.

Faber, J., Markowska, A., Nelissen, D., Davidson, M., Eyring, V., Cionni, I., Selstad, E., Kågeson, P., Lee, D., Buhaug, Ø., Lindtsad, H., Roche, P., Humpries, E., Graichen, J., Cames, M., & Schwarz, W. (2009). *Technical support for European action to reducing Greenhouse Gas Emissions from international maritime transport* (p. 353). CE Delft.

Faber, J., Nelissen, D., Hon, G., Wang, H., & Tsimplis, M. (2012). *Regulated slow steaming in maritime transport: An assessment of options, costs and benefits*. CE Delft.

Faber, J., Wang, H., Nelissen, D., Russell, B., & St Amand, D. (2011). Marginal abatement costs and cost effectiveness of energy-efficiency measures. In Organization, I.M. (Ed.), *The society of Naval Architects and Marine Engineers (SNAME)*. London.

Felício, J. A., Caldeirinha, V., & Dionísio, A. (2015). The effect of port and container terminal characteristics on terminal performance. *Maritime Economics & Logistics, 17*(4), 493–514.

Fontagné, L., & Fouré, J. (2013). Opening a Pandora's box: Modelling world trade patterns at the 2035 horizon. Paris, CEPII Working Paper 22.

Gilbert, P., Bows-Larkin, A., Mander, S., & Walsh, C. (2014). Technologies for the high seas: Meeting the climate challenge. *Carbon Management, 5*(4), 447–461.

Graham, S., Graham, B., & Holt, D. (2018). The relationship between downstream environmental logistics practices and performance. *International Journal of Production Economics, 196*, 356–365.

Ha, M. H., Yang, Z., Notteboom, T., Ng, A. K., & Heo, M. W. (2017). Revisiting port performance measurement: A hybrid multi-stakeholder framework for the modelling of port performance indicators. *Transportation Research Part E, 103*, 1–16.

Hanson, J., Brynoff, S., Fridell, E., & Lehtveer, M. (2020). The potential role of ammonia as marine fuel based on energy systems modelling and multi-criteria decision analysis. *Sustainability*, Forthcoming.

Heitmann, N., & Peterson, S. (2014). The potential contribution of the shipping sector to an efficient reduction of global carbon dioxide emissions. *Environmental Science & Policy, 42*, 56–66.

Hoegh-Guldberg, O., Caldeira, K., Chopin, T., Gaines, S., Haugan, P., Hemer, M., Howard, J., Konar, M., Krause-Jensen, D., Lindstad, E., Lovelock, C. E., Michelin, M., Nielsen, F. G., Northrop, E., Parker, R., Roy, J., Smith, T., Some, S., & Tyedmers, P. (2019). *The ocean as a solution to climate change: "Five opportunities for action." Report*. World Resource Institute.

Hoffmann, P. N., Eide, M. S., & Endresen, Ø. (2012). Effect of proposed CO2 emission reduction scenarios on capital expenditure. *Maritime Policy & Management, 39*(4), 443–460.

ICS. (2019). Proposal to establish an International Maritime Research and Development Board (IMRB) Submitted to IMO/MEPC 75 by BIMCO, CLIA, ICS, INTERCARGO, INTERFERRY, INTERTANKO, IPTA, and WSC.

Imai, A., Nishimura, E., Papadimitriou, S., & Liu, M. (2006). The economic viability of container mega-ships. *Transportation Research Part, E42*(1), 21–41.

IMO. (2009). *Second IMO greenhouse gas study 2014*. London: International Maritime Organization.

IMO. (2014). *Third IMO greenhouse gas study 2014*. London: International Maritime Organization.

IMO. (2018). Initial IMO Strategy on Reduction of GHG Emissions from Ships; IMO doc MEPC 72/17/Add.1, Annex 11, Resolution MEPC.304(72) (Adopted on 13 April 2018); IMO: London, UK, 2018.

IPPC - Intergovernmental Panel on Climate Change. (2000). Emission scenarios. http://sedac.ipcc-data.org/ddc/sres/ Nakicenovic, N. et al (2000). Special Report on Emissions Scenarios: A Special Report of Working Group III of the Intergovernmental Panel on Climate Change, Cambridge University Press, Cambridge, U.K., 599 pp. Available online at http://www.grida.no/climate/ipcc/emission/index.htm

ITF. (2019). ITF Transport Outlook 2019, International Transport Forum at the OECD, Paris.

Johansson, A., Guillemette, Y., Murtin, F., Turner, D., Nicoletti, G., de la Maisonneuve, C., Bagnoli, P., Bousquet, G., & Spinelli, F. (2012). Looking to 2060: Long-term global growth prospects. OECD Economic Policy Papers N°3, November.

Koesler, S., Achtnicht, M., & Köhler, J. (2020). Course set for a cap? A case study among ship operators on a maritime ETS. *Transport Policy, 37*, 20–30.

Kollamthodi, S., Pueyo, A., Gibson, G., Narkeviciute, R., Hawkes, A., Cesbron, S., Milnes, R., Harries, J., Zamparutti, T., Hernandez, G., Kaltsouni, S., Vancauwenbergh, S., Goldenman, G., Pålsson, C., Bengtsson, N., Rydbergh, T., Nilsson, L., Krantz, A., Weber, K., Scarbrough, T., Whall, C., Green, C., Hill, J., Lee, J., Noden, R., Grebot, B., & Lindstad, E. (2013). Support for the impact assessment of a proposal to address maritime transport greenhouse gas emissions, Report for European Commission – DG Climate Action. Ricardo-AEA/R/ED56985.

Leaper, R. (2019). The role of slower vessel speeds in reducing greenhouse gas emissions, underwater noise and collision risk to whales. *Frontiers in Maritime Science, 6*, 505.

Leimbach, M., Kriegler, E., Roming, N., & Schwanitz, J. (2017). Future growth patterns of world regions – A GDP scenario approach. *Global Environmental Change, 42*, 215–225.

Lindstad, E. (2013). *Strategies and measures for reducing maritimeCO2 emissions*. Doctoral thesis PhD. Norwegian University of Science and Technology. Department of Marine Technology.

Lindstad, E., Bø, T., & Eskeland, G. S. (2018). MARINE DESIGN XIII. In P. Kujala & L. Lu (Eds.), *Reducing GHG emissions in shipping- measures and options* (pp. 923–930). Taylor & Francis.

Lindstad, E., Borgen, H., Eskeland, G. S., Paalson, C., Psaraftis, H., & Turan, O. (2019). The need to amend IMO's EEDI to include a threshold for performance in waves (realistic sea conditions) to achieve the desired GHG reductions. *Sustainability, 2019*(11), 3668. https://doi.org/10.3390/su11133668

Lindstad, H., Asbjørnslett, B., & Strømman, A. (2011). Reductions in greenhouse gas emissions and cost by shipping at lower speeds. *Energy Policy, 39*, 3456–3464.

Lindstad H, Asbjørnslett BE, Jullumstrø E (2013). Assessment of profit, cost and emissions by varying speed as a function of sea conditions and freight market. Transportation Research Part D 19: 5-12.

Lindstad, H., Eskeland, G. S., Psaraftis, H., Sandaas, I., & Strømman, A. H. (2015a). Maritime shipping and emissions: A three-layered, damage-based approach. *Ocean Engineering, 110*, 94–101.

Lindstad, H., Verbeek, R., Blok, M., Zyl, S., Hübscher, A., Kramer, H., Purwanto, J., & Ivanova, O. (2015b). GHG emission reduction potential of EU-related maritime transport and on its impacts. Ref CLIMA.B.3/ETU/2013/0015. TNO report / TNO 2014 R11601 / 3. July 2015. Delft, The Netherlands.

McKinsey Global Institute. (2019). *Globalization in transition: The future of trade and value chains, January*. Retrieved at https://www.mckinsey.com/featured-insights/innovation-and-growth/globalization-in-transition-the-future-of-trade-and-value-chains

Meng Q, Du Y, Wang Y (2016). Shipping log data based container ship fuel efficiency modeling, Transportation Research Part B 83: 207–229.

Notteboom, T. E., & Vernimmen, B. (2009). The effect of high fuel costs on liner service configuration in container shipping. *Journal of Transport Geography, 17*(5), 325–337.

OECD. (2018). The long view: Scenarios for the World Economy to 2060. OECD Economic Paper No. 22, 50 p.

OECD. (2019). Decarbonisation maritime transport: Pathways to zero-carbon shipping by 2035. OECD/International Transport Forum.

Offshore Energy. (2020). Major global traders commit to reporting on shipping emissions. Available at (12 October 2020): https://www.offshore-energy.biz/major-global-traders-commit-to-reporting-on-shipping-emissions/?utm_source=worldmaritimenews&utm_medium=email&utm_campaign=newsletter_2020-10-08

Olmer, N., Comer, B., Roy, B., Mao, X., & Rutherford, D. (2017). *Greenhouse gas emissions from global shipping, 2013-2015*. International Council on Clean Transportation, 28 p.

Panagakos, G., Pessôa, T. S., Dessypris, N., Barfod, M. B., & Psaraftis, H. N. (2019). Monitoring the carbon footprint of dry bulk shipping in the EU: An early assessment of the MRV regulation. *Sustainability, 11*, 5133. https://doi.org/10.3390/su11185133

Pazirandeh, A., & Jafari, H. (2013). Making sense of green logistics. *International Journal of Production Performance and Management, 62*(8), 889–904.

Pereira, C. (2019a). EEOI & AER & ISPI Evaluation, Thetis-MRV Database. European Sustainable Shipping Forum (ESSF), Brussel, Belgium 21. October 2019.

Pereira, C. (2019b). Energy Efficiency Analysis, Thetis-MRV Database. European Sustainable Shipping Forum (ESSF), Brussel, Belgium 18. December 2019.

Pereira, C. (2020). Technical Efficiency vs. Operational Performance, Thetis-MRV- Evaluation. European Sustainable Shipping Forum (ESSF), 3. March 2020.

Plum, C. E. M., Pisinger, D., Salazar-González, J. J., & Sigurd, M. (2014). Single liner shipping service design. *Computers & Operations Research, 45*, 1–6.

Psaraftis, H. N. (2016). Green maritime logistics: The quest for win-win solutions, 6th Transport Research Arena, April 18–21. *Transportation Research Procedia*, 133–142.

Psaraftis, H. N. (2019). Speed optimization vs speed reduction: The choice between speed limits and a bunker levy. *Sustainability, 11*(8), 22–49.

Psaraftis HN, Kontovas CA (2013). Speed models for energy-efficient maritime transportation: A taxonomy and survey, Transportation Research Part C 26: 331-351.

Psaraftis HN, Kontovas CA (2014). Ship speed optimization: Concepts, models and combined speed-routing scenarios, Transportation Research Part C 44: 52-69.

Raucci, C., Prakash, V., Rojon, I., Smith, T., Rehmatulla, N., & Mitchel, J. (2017). *Navigating decarbonisation: An approach to evaluate shipping's risks and opportunities associated with climate change mitigation policy*. Commissioned by Carbon War Room (CWR), UMAS, February 2017.

Smith, T., Jalkanen, J., Anderson, B., Corbett, J., Faber, J., Hanayama, S., O'Kee, E., Parker, S., Johansson, L., & Aldous, L. (2014). *Third IMO GHG Study 2014*. IMO.

Stott, P., & Wright, P. (2011). Opportunities for improved efficiency and reduced CO2 emissions in dry bulk shipping stemming from the relaxation of the Panamax beam constraint. *International Journal of Maritime Engineering, 153*(A4), A215–A229.

Tang, S., Wang, W., Yan, H., & Hao, G. (2015). Low carbon logistics: Reducing shipment frequency to cut carbon emissions. *International Journal of Production Economics, 164*, 339–350.

Tillig, F., Mao, W., & Ringsberg, J. (2015). *Systems modelling for energy-efficient shipping*. Chalmers University of Technology.

Tran, N. K., & Haasis, H. D. (2015). An empirical study of fleet expansion and growth of ship size in container liner shipping. *International Journal of Production Economics, 159*, 241–253.

Traut, M., Larkin, A., Anderson, K., McGlade, C., Sharimina, M., & Smith, T. (2018). CO_2 abatement goals for international shipping. *Climate Policy, 18*(8), 1066–1075.

UMAS. (2018). Zero-Emission Vessels 2030. How do we get there? Low Carbon Pathways 2050 Series.

UMAS. (2020). *Techno-economic assessment of zero-carbon fuels*. Retrieved at: https://www.lr.org/en/latest-news/lr-and-umas-publish-techno-economic-assessment-of-zero-carbon-fuels/

UNCTAD. (2017). Review of maritime transport. UNCTAD/RMT/2017.

UNCTAD. (2018a). *50 years of review of maritime transport 1968-2018: Reflecting on the past, exploring the future*. UNCTAD publication. https://unctad.org/en/PublicationsLibrary/dtl2018d1_en.pdf

UNCTAD. (2018b). Review of maritime transport. UNCTAD/RMT/2018.

van Vuuren, D. P., Kriegler, E., O'Neill, B. C., Ebi, K. L., Riahi, K., Carter, T. R., Edmonds, J., Hallegatte, S., Kram, T., Mathur, R., & Winkler, H. (2014). A new scenario framework for climate change research: Scenario matrix architecture. *Climatic Change, 122*, 373–386.

van Vuuren, D. P., Stehfesta, E., Gernaata, D. E. H. J., Doelman, J. C., van den Berga, M., Harmsena, M., de Boer, H. S., Bouwmana, L. F., Daioglou, V., Edelenbosch, O. Y., Girod, B., Krama, T., Lassaletta, L., Lucas, P. L., van Meijle, H., Müller, C., van Ruijven, B. J., van der Sluisa, S., & Tabeau, A. (2017). Energy, land-use and greenhouse gas emissions trajectories under a green growth paradigm. *Global Environmental Change, 42*, 237–250.

Wang, H., & Lutsey, N. (2013). Long-term potential for increased shipping efficiency through the adoption of industry-leading practices. In: *International Council on Clean Transportation*, September 30th.

Wang, S., Ji, B., Zhao, J., Liu, W., & Xu, T. (2018). Predicting ship fuel consumption based on LASSO regression. *Transportation Research Part D, 65*, 817–824.

Wang, S., & Meng, Q. (2012). Sailing speed optimization for container ships in a liner shipping network. *Transportation Research Part E, 48*, 701–714.

Wärtsila. (2009). Boosting energy efficiency: Energy efficiency catalogue. In *Energy Efficiency Catalogue/Ship Power R&D*. Wärtsila.

Green Shipping Finance: Existing Initiatives and the Road Ahead

Manolis G. Kavussanos and Dimitris A. Tsouknidis

Abstract The transportation service provided by the global shipping industry contributes substantially to the phenomenon of climate change through vessel greenhouse-gas (GHG) emissions. This paper discusses the developing regulatory framework on green shipping finance and the role of capital providers in reducing the environmental footprint of the industry. Several initiatives by regulatory bodies and private companies are examined, whose purpose is to reduce shipping-related emissions through green financing schemes. The paper also provides a critical discussion on the progress so far and the road ahead.

1 Introduction

Climate change as a result of greenhouse-gas (GHG) emissions constitutes a global (negative) externality—a public "bad" as opposed to a public good—and highlights the inability of private markets to provide an efficient allocation of resources (Nordhaus, 2019). Human activities related with climate change spill their costs outside the market and are typically not captured in market prices. Being a global (negative) externality, climate change needs a different treatment rather than a local or national public good (or bad), as it is out of the control of the markets and national governments (Nordhaus, 2019). Thus, it seems that climate change requires the enforcement of cooperative international policies, which are often not easy to be implemented, since individual nations only enjoy a small percentage of their compliance actions. An industry of special attention is the global shipping industry, since it exhibits a substantial environmental footprint.

Vessels burn heavy fuel oil and marine diesel oil. This produces both greenhouse gas (GHG) and non-GHG emissions. GHG emissions include carbon dioxide (CO_2), methane (CH_4) and nitrous oxide (N_2O). GHG emissions are directly related to ships' fuel consumption, which in turn depends on a number of factors, including the

M. G. Kavussanos (✉) · D. A. Tsouknidis
Athens University of Economics and Business, Athens, Greece
e-mail: mkavus@aueb.gr

B.-W. Ko, D.-W. Song (eds.), *New Maritime Business*, WMU Studies in Maritime Affairs 10, https://doi.org/10.1007/978-3-030-78957-2_6

vessel's hull shape and roughness, loading conditions, engine condition, and weather conditions amongst others. The main emissions and air pollutants of concern are Sulphur Oxides (SOx), which are harmful to human health and the environment.

The global shipping industry carries over 80% of world trade in volume terms according to research by the United Nations Conference on Trade And Development (UNCTAD, 2019) and emits around 940 million tonnes of CO_2 annually. This represents about 2.5% of global greenhouse gas (GHG) emissions according to the third International Maritime Organization (IMO) GHG study (IMO, 2014).[1] Unlike the sectors of energy, industrial processes, agriculture and waste, the shipping industry has not been under specific obligations or reduction targets regarding GHG emissions, imposed by the signing of the Kyoto Protocol in 1997.[2] Also, the Paris agreement, signed in 2015, made no reference to the sector of international shipping.

In 2011, the IMO, being the main regulatory body responsible for the shipping industry, amended Annex VI of the International Convention for the Prevention of Pollution from Ships (MARPOL) and introduced new mandatory standards through two technical measures: the Energy Efficiency Design Index (EEDI) and the Ship Energy Efficiency Management Plan (SEEMP). These measures do not set strict sector-wide emission targets but help in achieving less emissions through measuring the efficiency of ships; leading to less fuel consumption and eventually less emissions. Specifically, the EEDI introduces minimum efficiency design requirements for all new ships; while SEEMP defines a technological threshold for all ships to meet energy efficiency requirements, such as for example less fuel consumption (Resolution MEPC.203.62).[3]

In 2018, the IMO announced the first shipping industry-wide emissions reduction target, which aims to reduce emissions by at least 50% by 2050 compared to the 2008 level. Several measures have been proposed by IMO in its initial strategy, the so-called IMO Strategy on Reduction of GHG Emissions from Ships, to achieve this

[1]Source: URL (last access 6th of September 2020): http://www.imo.org/en/OurWork/Environment/PollutionPrevention/AirPollution/Pages/Greenhouse-Gas-Studies-2014.aspx. According to IMO: "Based on the recommendations from an Expert Workshop held in March 2019, the Marine Environment Protection Committee (MEPC 74) approved the terms of reference of the Fourth IMO GHG Study, submitted to MEPC 76 in Autumn 2020."

[2]The Kyoto Protocol, adopted in 1997 and entered into force for its 192 parties (states) in early 2005, is an international treaty extending the 1992 UN Framework Convention on Climate Change (UNFCCC), which commits participating countries to reduce GHG emissions.

[3]On 20th of November 2020 new mandatory regulations to cut the carbon intensity of existing ships were approved by IMO's Marine Environment Protection Committee (MEPC), which are in line with the IMO's (2018) strategy on GHG emissions and are expected to be enforced by 2023. Specifically, a new Energy Efficiency Existing Ship Index (EEXI) is introduced whose standards will be largely equivalent to the current Energy Efficiency Design Index (EEDI) for newbuild ships in 2020. Furthermore, a new operational Carbon Intensity Indicator (CII) will be used to rank vessels in terms of fuel efficiency (ranging from A and E, with ships rated D and below falling under the minimum standard). These measures aim to address both the technical and operational aspects of vessels' environmental performance.

target—see among others Garcia et al. (2020). Thus, technical and operational energy efficiency measures were proposed through the IMO Strategy on Reduction of GHG Emissions from Ships. The focus was on improving EEDI and SEEMP frameworks, as well as technical cooperation and capacity building. The measures and time-plans can be summarized as follows: (a) National Action Plans, which are deemed short-term and span the period 2018–2023, aiming to create supportive domestic policies; (b) Market-based mechanisms, deemed as medium-term and cover the period 2023–2030. These aim to work through the introduction of market (tax) incentives for cleaner fuels; (c) The use of alternative fuels to replace polluting ones over a long-term strategy extending beyond 2030. The objective is to shift resources and capital towards low or zero-carbon fuels by 2030.

The voluntary and mandatory measures taken primarily by the IMO to reduce the environmental footprint of the international shipping industry can have significant repercussions for various participants involved in the global maritime industry, including the shipping finance segment of it. To make the point, it is important to realize that the growing interest in the reduction of the environmental footprint of international shipping creates considerable pressure for shipping companies, charterers, ship-lending banks and institutional investors to treat and manage climate change as a financial risk. IMO's initiatives combined with increased pressure by charterers and eventually consumers for low-carbon shipping options, impose a considerable cost of compliance for shipping companies. Furthermore, the necessary reduction of the use of fossil fuels due to their high carbon intensity, such as coal and oil, directly affects the demand for the vessels that carry them. Thus, it is likely that environmental regulations will result in less employment opportunities for a large part of the global fleet, such as Capesize and Panamax vessels in the dry bulk sector and tanker vessels carrying crude oil and its products in the liquid bulk sector.

Such business risks are of primary concern to financial institutions, institutional and private investors who are also exposed to these types of risks through their portfolio of shipping bank loans and/or their equity/debt investments in the shipping sector. For instance, ship-lending banks may be facing an increase in shipping bank loan defaults due to a deterioration of the cash-flow generating ability of shipping firms which operate vessels in the aforementioned segments of the shipping industry. The same is also true for debt and equity holders in shipping companies that remain exposed to 'green-obsolete' projects; that is, to vessels that in one way or another are not emission-friendly to the environment and become economically obsolete, either because they are involved in the carriage of 'non-green' commodities or they are not 'green' to operate.

In addition, the potential ship technological obsolescence as well as the reduction in the cash flow generating ability of these assets will lead to lower asset values. Typically, these vessels are used as collateral in bank/bond loan agreements. A reduction in their value, due to their lower cash flow generating ability, implies diminishing the value of the main collateral of a lending agreement, being a bank loan or a bond issue. This increases the probability of the credit facility remaining uncovered in case of default. Moreover, vessels are the main assets in company valuations, and a reduction in their values constitutes a business risk for the investors in these companies.[4]

Despite the above risks, these regulations can be thought of as an opportunity to allocate capital resources in a way that contributes towards achieving the global emission reduction targets. A promising policy tool to incentivize shipping market players to adopt and comply with the low carbon footprint target for international shipping is the terms and conditions of financing low carbon (*green*) ocean-going vessels, as will be discussed later in the paper. The role of capital providers in incentivizing the transition to green ocean-going vessels, through financing eco-friendly newbuildings or retrofitted second-hand vessels, can be significant.

This paper contributes to the extant literature by providing a detailed and critical discussion of the timely and pressing question of the role of capital providers in reducing the environmental footprint of the international ocean-going shipping industry. In the process, the paper reviews and synthesizes the public and private initiatives towards the financing of sustainable (green) investments in the shipping sector. Furthermore, the paper places green shipping finance as a sub-topic within the general literature on shipping finance and critically discusses the progress so far and the road ahead. The rest of this paper is organized as follows: Sect. 2 reviews the public and private initiatives in the green shipping finance field, while Sect. 3 concludes the paper.

2 Financing Schemes for Green Shipping

This section of the paper provides a review of the various ship financing schemes related to green shipping. They involve initiatives by both supranational bodies and private entities in allocating resources towards reducing the environmental footprint of the global maritime industry.

2.1 *Taskforce on Climate-Related Financial Disclosures (TCFD) of the Financial Stability Board (FSB)*

A voluntary initiative taken from the Financial Stability Board in 2017 is the Taskforce on Climate-related Financial Disclosures (TCFD), under which it is

[4]For a full review of the literature on shipping finance see Kavussanos and Visvikis (2016) and Alexandridis et al. (2018); while credit risk related empirical investigations in the shipping finance literature are provided by Kavussanos and Tsouknidis (2014, 2016).

recommended for business and investors to conduct scenario analysis to assess the financial impact of climate change and adjust their business strategy accordingly.[5]

A recommendation of special interest from TCFD is that firms should disclose climate-related financial disclosures in their annual financial filings. The TCFD believes that including material climate-related information in firms' filings will foster shareholder engagement and broader use of climate-related financial disclosures. Eventually, this is expected to promote a better understanding of climate-related risks and opportunities by investors and other users of financial statements. Furthermore, the mandatory disclosure of climate-related information will set in place the appropriate mechanisms to monitor and govern the accuracy of the disclosed information, i.e. through the chief financial officer, the audit committee and external evaluators, also in accordance with national disclosure requirements. Overall, TCFD's voluntary recommendations aim to provide a common international framework for the disclosure of climate change related information of companies, banks, insurers and investors. The recommendations are classified into four broad categories, namely governance, strategy, risk management and metrics.

It is important to note at this point that the Association of the Principles for Responsible Investing (PRI) (based in the UK) supports the adoption of the recommendations of TCFD, stating that 'they provide a global framework for translating information about climate into financial metrics'. The PRI Association, with signatories, asset owners, investment managers and service providers, introduced TCFD-aligned indicators to its Reporting Framework in 2018. This is a very important development as the Association involves over 480 investors representing US$42 trillion, who opted to complete the indicators and submit responses.

From 2020, the climate indicators are mandatory to report but voluntary to disclose. The remaining PRI climate risk indicators will stay voluntary with a view to becoming mandatory as good practice develops. The PRI is concerned that the longer the world goes without a safe trajectory on climate change, the greater the risks for investors of an abrupt policy response. They state that 'It is thought important for investors to incorporate emerging mega risks such as climate change into their view of the future. TCFD provides the best available framework for systematically including climate-related risks and opportunities into investment strategy.'

However, the impact of TCFD's recommendations on the firms raising funds through bank loans may vary across industries. For example, the impact of the disclosure of climate related information on industries known to operate under conditions of almost perfect competition, such as the global shipping industry, may be different compared to other industries with less impact on the environment.

Overall, financial institutions, and in particular ship-lending banks, can help substantially in achieving emission reduction targets. This is related to their primary

[5]For details see, "TCFD-based reporting to become mandatory for PRI signatories in 2020" at the following URL (last access 14th September 2020): https://www.unpri.org/news-and-press/tcfd-based-reporting-to-become-mandatory-for-pri-signatories-in-2020/4116.article.

function, which is to assess the business viability of the most promising investments and allocate capital accordingly. For decision purposes, taking into account during the evaluation process climate related factors can lead to allocation of resources into 'greener' projects.

2.2 *Institutional Initiatives: The EU and the European Investment Bank (EIB) Initiatives*

In line with the earlier discussed rationale, the EU itself has developed a number of policies to enhance the role of financial institutions in financing sustainable investments across sectors, which includes the maritime industry. Specifically, in 2018, the European Commission disclosed an Action Plan on Sustainable Finance, where it develops a number of strategies towards sustainable finance and classifies (providing a taxonomy) economic activities according to their contribution towards sustainability objectives, including climate change.[6] Sustainability criteria have been developed for land transport and are likely to include maritime shipping in the near future, including zero emission fleets and efficiency measures.

Taking action in supporting green finance in the shipping industry is of particular importance at a global level. This is because (1) investing in greener technologies represents an upfront cost for shipowners, possibly rendering the ones that comply at the outset of the measures less competitive and (2) access to shipping finance has become overall more difficult after 2008 due to several withdrawals/defaults of shipping market players in the years after the global financial crisis and the onset of Basel criteria in the evaluation of the riskiness of bank lending portfolios. Specifically, according to Basel IV (2016) criteria, banks are required to hold more reserve capital, thereby reducing their liquidity, for riskier projects in their portfolios; shipping loans fall in this category. Thus, it has been observed that particularly EU and US banks have reduced substantially their shipping credit portfolios after the 2008 financial-economic-shipping crisis.

In order to practically support greener shipping investments, the European Investment Bank's (EIB) Board of Directors approved its €750 million Green Shipping Financing Programme in June 2016, also known as the Green Shipping Guarantee Programme, GSG. The funds are provided by the Connecting Europe Facility-CEF (€250 million) and the European Fund for Strategic Investments EFSI (€500 million). Specifically, the GSG programme is designed to promote commercial investment in existing and upcoming technologies, which will enhance fuel/energy efficiency and reduce GHG emissions in the European shipping sector.

[6]This Action Plan also introduced a number of changes to earlier regulation on carbon impact benchmarks, see EU Regulation 2019/2089, available at (last access 23rd October 2020): https://eur-lex.europa.eu/legal-content/EN/TXT/PDF/?uri=CELEX:32019R2089&from=EN.

EIB's initiative seeks to address market gaps and constraints to access finance within the shipping sector and make ship-lending attractive for commercial banks. This is achieved by reducing the credit risk of shipping bank loans for environmentally focused investments. Specifically, EIB reviews over time the main financing constraints in shipping finance, the larger being the risks associated with the sector but also the reluctance of banks to provide the capital needed for environmentally focused investments, especially on the existing fleet. Therefore, EIB's GSG Programme aims at mitigating credit risk of credit facilities granted to environmentally friendly investments within the shipping industry. The programme is developed in cooperation with ship-lending financial institutions, in order to provide advantageous financial terms to green shipping investments, backed up by the EIB's AAA credit rating. The EIB supports up to 100% of the incremental environmental investments on existing vessels or up to 50% of the debt financing on new vessels. In both cases the overall financing will be between 20% and 50% of the size of the total shipping bank loan. The remaining of the loan will be provided by commercial banks, leading to a co-financing structure for these credit facilities. Therefore, the financing of the GSG programme is provided by a Partner Financial Institution (PFI).[7] Specifically, the individual loan agreements are initiated and assessed by the PFI and are provided to EIB for internal review, pricing and approval. Loans involving up to €50 million EIB's contribution undertake a streamlined approval. This co-financing structure of the loans provided, in a way ensures that ship operators without a 'bankable' business plan will not be supported through EIB platform for green shipping financing. In this way, the inducement of a distortion effect in the market is avoided.

A sister programme of green ship financing initiated also by EIB, is the Green Shipping Programme Loan, which forms a direct individual EIB loan to the borrower in order to enable investment in green shipping. Eligible investments are available for shipowners with significant European interests (flag, ownership, routes) and may include: shipbuilding of new green vessels; environmentally friendly retrofittings in existing vessels, such as hull treatment, ballast water treatment, etc., and; promoting the adoption of the use of alternative fuels, such as LNG. Funds available are up to €250 million provided by the European Fund for Strategic Investments (EFSI). In this case, the potential borrower (typically a shipowner/ship operator) applies directly to the EIB for financing, through EIB's local offices. The programme may fund up to 50% of debt financing. However, EIB's green shipping finance initiatives do not aim to change the structural issues of the sector nor will provide finance to non-eligible investments and/or ship operators not presenting a sound business plan.

[7]PFI's participating in the GSG Programme include: Societe Generale, BNP Paribas, ANB AMRO, ING, Credit Agricole, etc. For more details see the following URL (last access 30th November 2020): https://www.shortsea.gr/wp-content/uploads/2019/06/Maria-Tzoumanika-European-Investment-Bank.pdf.

According to EIB, loans provided by either the Green Shipping Guarantee Programme or the Green Shipping Programme Loan need to comply with specific eligibility criteria, such as: (1) The eligibility under the EIB Transport Lending Policy with significant European interest (flag, ownership, routes); (2) funds need to be intended to finance both new vessels and retrofittings, that promote sustainable transport and environmental protection, including Climate Action initiatives; (3) there needs to be a particular focus on green investments, such as dual fuel engine technology, exhaust gas cleaning system (scrubbers), propulsion technology, hull treatment, ballast water treatment systems, etc.; (4) not prescribing specific technologies, in order to be eligible any technology proposed should be able to demonstrate a significant reduction in harmful air borne emissions or water borne discharges.

The full assessment of the scope of each candidate project to be financed is performed during its corresponding appraisal. The shipowners and shipyards financed through this programme are expected to be well established (experienced) and secure all the relevant environmental certifications for their vessels conforming fully with the EU / European Maritime Safety Agency (EMSA) and the IMO international maritime conventions. Therefore, building new vessels or retrofitting existing ones through this co-financing programme will need to be certified by an International Association of Classification Societies (IACS) member or an EU / EMSA recognized classification society.

According to EIB's documentation, the objective of the green shipping finance initiatives of EIB extend beyond environmentally friendly newbuilding vessels and retrofitting of existing vessels. For example, the EIB will seek to support through attractive financing terms those shipowner/ship operators who are able to exhibit their commitment and plans towards further enhancing their environmental and social awareness, processes and systems in place. This may be achieved for example through adopting other non-regulatory international standards, such as the ISO standards related to the shipping industry. These may include, among others: ISO9001 Quality Management, ISO14001 Environmental Management, ISO47 Shipbuilding and Marine Structures, OHSAS18001 Occupational Health and Safety, etc.

On 8 January 2020, EIB announced the financing of the construction of a new liquefied natural gas (LNG) bunkering vessel, through its flagship GSG Programme, to be based at Piraeus Port (Core TEN-T port). The owner of the LNG bunkering vessel is the Greek national provider of natural gas, DEPA. The total cost of constructing the vessel has been announced to be approximately €40 million, with EIB's proposed contribution standing at approximately €20 million. The vessel's core operation will be delivering gas to LNG-fuelled vessels (ship-to-ship bunkering), mainly within the port of Piraeus and in other smaller neighbouring ports in the Aegean and East Mediterranean Sea. The vessel under construction complies with the EU and IMO regulations and will operate under a EU Member State flag. The aim of the investment is to provide a mobile LNG bunkering option and promoting LNG as an alternative fuel source for the shipping industry. In this way, the financing of this vessel is expected to reduce emissions from maritime transport in the wider region, thus reducing the overall environmental impact.[8]

2.3 *Private Initiatives: The Poseidon Principles initiative*

There are also other (private) initiatives aiming to support financially green projects in shipping and direct resources towards them, in alignment with IMO's decarbonization targets; the aim is to allow shipping companies with these characteristics to access finance in favourable terms. The 'Poseidon Principles' is an initiative launched in June 2019, where eighteen (18) leading signatory banks[9] agreed to incorporate climate change criteria into their lending decisions and in this way enhance sustainable shipping investments. Specifically, participant banks commit to measure and report annually the carbon intensity of their shipping credit portfolios against IMO's decarbonization strategy target; that is, reducing the total annual GHG emissions by at least 50% by 2050 compared to the 2008 levels. In this way, the signatory banks will use the carbon intensity of a vessel, product or portfolio relative to established decarbonization trajectories to measure climate alignment. The main metric used is the Annual Efficiency Ratio (AER), which uses the parameters of fuel consumption, distance travelled, and deadweight tonnage at summer draught. The final output is a decarbonization trajectory, which is a time-series representation of how many grams of CO_2 a single ship emits to move one tonne of goods one nautical mile over a time horizon. Standard decarbonization trajectories are produced by the Secretariat of the Poseidon Principles for each ship type and size class.

The Poseidon Principles introduce four core principles, which are discussed next. (1) Assessment of Climate Alignment: Each signatory member commits itself in measuring the carbon intensity of their shipping credit portfolio on an annual basis and comparing this metric to the decarbonisation trajectory set by the IMO. (2) Accountability: signatory banks commit to follow the data types, sources and service providers established by the IMO's Fuel Oil Data Collection Service (DCS). The DCS requires owners to collect, and report to the flag state, data on fuel types and consumption and it is mandatory for ships over 5000 gross tonnage. Then, the flag state assesses whether the data reported comply with the IMO requirements and issues the statement of compliance for the vessel. (3) Enforcement: signatories are strongly recommended to include documentation of the standardised covenant

[8]More details on the financing of this specific LNG bunkering vessel may be found in the following URL (last access 30th November 2020): https://www.eib.org/en/projects/pipelines/all/20190313.

[9]The 18 'Poseidon Principles' participating banks represent jointly US$150 billion of investments in shipping finance. They include ABN-AMRO, Amsterdam Trade Bank, BNP Paribas, bpi france, CIC, Citi, Credit Agricole, Credit Suisse, Danish Ship Finance, Danske Bank, DNB, DVB, EKSPORTKREDITT, ING, Nordea, Societe Generale, Sparebanken Vest, Sumitomo Mitsui Trust Bank.

clause, which is explained in detail in the Poseidon Principles documentation, and requires from the shipowners/operators to provide the appropriate data regarding vessels' consumption in order to calculate vessels' carbon intensity. This information is also mandatory to be provided in the DCS. (4) Transparency: signatories must acknowledge publicly that they are members of the Poseidon Principles. Furthermore, every year, signatories need to collect from the shipowners/operators the relevant data submitted to the DCS and calculate the carbon intensity for each individual vessel and the total shipping credit portfolio. Then, the credit portfolio climate scores are assessed vs. the objective trajectory set by IMO and reported by the end of December by the Poseidon Principles Association. An important implication of the procedure described above is that under the Poseidon Principles, signatories agree to include a not compulsory, but strongly recommended, standardised covenant which includes exact definitions of the information needed to be reported by the shipowner. However, ship-lending banks are also able to obtain the needed climate-related information under the standard information covenants included in the syndicated ship finance documentation.

2.4 *Private Initiatives: The Green Loan Principles*

Another initiative is the 'Green Loan Principles' (GLP) initiative launched in 2018 by the Loan Market Association (LMA) of Europe, Middle East and African (EMEA)[10] countries, the Asia Pacific Loan Market Association (APLMA)[11] and the Loan Syndications and Trading Association (LSTA)[12] of North America. GLP is a set of recommended guidelines (criteria) to be applied by market participants on a voluntary basis to classify a shipping bank loan agreement as "green". According to the documentation of the GLP initiative, the borrower of a green loan should clearly communicate to its lenders the following information: (1) its environmental sustainability objectives; (2) the process by which the borrower determines how its project fits within the eligible categories set out in Appendix 1 of the GLP documentation; and (3) related eligibility criteria as set out in GLP documentation.[13] Shipowners/ship operators are encouraged to position this information within the context of their overarching objectives. For example their objectives regarding their strategy, policy

[10]Europe, Middle East, and Africa (EMEA) countries are a geographical division used by many multinational corporations. The acronym is popular among North American companies.

[11]APLMA is a professional not-for-profit trade association which represents the interests of institutions active in the syndicated loan markets in the Asia-Pacific ("APAC") region.

[12]Loan Syndications and Trading Association (the "LSTA") is a financial services trade group which exists to enhance the development and running of the North American syndicated loan market.

[13]The GLP documentation is available in the following URL (last access 24th October 2020): https://www.lma.eu.com/application/files/9115/4452/5458/741_LM_Green_Loan_Principles_Booklet_V8.pdf.

and/or processes relating to environmental sustainability. Furthermore, they are encouraged to disclose the green standards or certifications they aim to obtain and sustain.

The Categories of Eligibility for Green Projects under the GLP initiative include (indicative but not limited to):

- clean transportation—such as electric, hybrid, public, rail, non-motorised, multi-modal transportation, infrastructure for clean energy vehicles and reduction of harmful emissions;
- renewable energy—including production, transmission, appliances and products; energy efficiency—such as in new and refurbished buildings, energy storage, district heating, smart grids, appliances and products;
- pollution prevention and control—including reduction of air emissions, greenhouse gas control, soil remediation, waste prevention, waste reduction, waste recycling and energy/emission-efficient waste to energy;
- environmentally sustainable management of living natural resources and land use; climate smart farm inputs such as biological crop protection or drip-irrigation;
- environmentally sustainable fishery and aquaculture, environmentally sustainable forestry, including afforestation and reforestation, and preservation or restoration of natural landscapes;
- terrestrial and aquatic biodiversity conservation—including the protection of coastal, marine and watershed environments;
- sustainable water and wastewater management—including sustainable infrastructure for clean and/or drinking water, wastewater treatment, sustainable urban drainage systems and river training and other forms of flooding mitigation;
- climate change adaptation—including information support systems, such as climate observation and early warning systems;
- eco-efficient and/or circular economy adapted products, production technologies and processes—such as development and introduction of environmentally sustainable products, with an eco-label or environmental certification, resource-efficient packaging and distribution; and
- green buildings which meet regional, national or internationally recognized standards or certifications.

2.5 *Private Initiatives: The Getting to Zero Coalition Initiative*

Finally, strategic business coalitions have also started building momentum. For instance, the 'Getting to Zero Coalition' initiative, launched in June 2019, brings together decision-makers from across the shipping value chain with key stakeholders from the energy sector as well as from governments and intergovernmental organizations. It is a partnership between the Global Maritime Forum, the Friends of Ocean action, and the World Economic Forum. More than 120 companies are involved in this initiative representing the shipping, energy, supply chain and finance sectors.[14] The coalition supports the Decarbonization initiative launched in October 2018. Further knowledge support is provided by the University College (UCL) Energy Institute, the Environmental Defense Fund, and the Energy Transitions Commission.

The Coalition focuses on the development of commercially viable zero emission vessels and the associated fuel supply chain which must enter the fleet by 2030, in order to reduce the shipping sector GHG emissions by at least 50% by 2050, as set out by the IMO. It recognizes the fact that commercial vessels can be operated for 20 years or more. Therefore, ships constructed around the year 2030 are expected to be still in operation in 2050. At the same time, the required infrastructure related to fuel supply chains is expected to have an economic life of up to 50 years; while any attempted reconfiguration to new fuels can be a lengthy process. As a result, there is a pressing need to develop safe, technically feasible, commercially viable and zero-emission vessels to be constructed by 2030.

2.6 *Green Bonds*

The use of green bond issues (also known as climate bonds) forms a prominent alternative to green bank loans for shipping investments. Green bonds are fixed income securities to finance low carbon investments. The European Investment Bank (EIB) issued an equity index-linked bond in 2007, which became the first fixed income product linked to socially responsible investments.[15] Bonds structured in a "Climate Awareness Bond" structure were used to provide capital towards renewable energy and energy efficiency projects. Green bonds form a relatively new asset class but they are growing rapidly, as their total volume grew from US$ 160 billion in 2016 to US$ 255 billion in 2019.[16] Over time a large number of green bonds have been issued by several issuers globally, including governments, financial institutions and private companies.

Green bonds enable pension, government, insurance and sovereign wealth funds to invest in areas perceived as important to their stakeholders that have the same credit risk and return profiles as standards bonds. They also provide a way for governments to fund climate change mitigation, through, for example, preferential tax treatments. Apart from the scope of the proceeds through a green bond issue, which need to be environmentally friendly, green bonds exhibit largely the operational characteristics of conventional debt instruments. For example, they are credit

---

[14]For more details see the following URL (last access 18th September 2020): https://www.globalmaritimeforum.org/getting-to-zero-coalition.

[15]Details on EIB's green bond issue may be found in the following URL (Last access: 20th November 2020): https://ec.europa.eu/commission/presscorner/detail/en/BEI_13_109.

[16]The source of these data is in the following URL (Last access: 20th November 2020): https://www.climatebonds.net/2020/01/record-2019-gb-issuance-255bn-eu-largest-market-us-china-france-lead-top-20-national.

rated based on the creditworthiness of the issuer and tradable in international bond markets.

A controversial issue regarding green bonds that has attracted international criticism is the lack of a specific and universal framework defining when a bond issue qualifies to be labelled "green". However, the OECD (2016) has provided a definition of green bonds which mitigates to an extent the lack of well-defined criteria regarding being classified as "green". OECD's definition is as follows: "*Green bonds are debt instruments used to finance green projects that deliver environmental benefits. A green bond is differentiated from a regular bond by its commitment to use the funds raised to finance or refinance "green" projects, assets or business activities. Green bonds can be issued by either public or private actors up front to raise capital for projects or for refinancing purposes, freeing up capital and leading to increased lending.*"

Therefore, given the relatively generic definition of a green bond, it lies within investors (potential bondholders) to evaluate the purpose of the bond issuance, the allocation of the proceeds raised and the ways to monitor the progress of the green projects over time. Thus, a potentially important obstacle to the high potential of green bonds as a way to reduce the environmental footprint of business activities is the lack of a rigorous framework to ensure transparency and disclosure of information which would facilitate the interactions between investors and issuers and increase the credibility of the process.

The lack of a common framework has been addressed by the International Capital Market Association (ICMA), which in 2014 issued the "Green Bond Principles (GPB)", the first version of a set of voluntary guidelines for green bonds, which has been revised since then as in ICMA (2017). Green bonds follow the GBP's stated by ICMA and the proceeds from their issuance are used for the pre-specified types of projects. The GBP's have been widely used and gained recognition over the years, despite the fact that other authorities also issued similar principles for green bonds, such as for example the People's Bank of China, the Central Bank of China and the Japanese Ministry of the Environment.

2.6.1 Climate Bonds Initiative (CBI)

The London-based Climate Bonds Initiative (CBI) has provided the world's first Certification program for climate bonds, which has been used as a blueprint for various countries to set up their own criteria for labelling green bonds. These CBI criteria are expected to function as a universal and credible set of rules for green investments in the shipping industry. According to CBI: "*The criteria require that ships are not primarily dedicated to transporting fossil fuels and are either zero-emission ships at the point of application for certification or performing below the average operational emissions intensity of their respective size and class throughout the tenor of the bond.*" Furthermore, CBI's guidelines stress that ships applying for certification, i.e. inclusion in the green bond category, need to perform better rather than the average of their group and must also draft an emission reduction plan on their continued effort to transit into zero emissions up to 2050.

CBI's green bond criteria form the most recent initiative to enhance green financing within the shipping industry. The rationale of this initiative is to reward the sustainability-minded shipowners and operators by transmitting a signal to potential investors that the specific shipping company has a clear plan to mitigate its emissions footprint. CBI's criteria for labelling a bond issue as green, provide higher transparency and credibility to the market. In this way, CBI requires potential green bond issuers to consider how their investment, the vessel in the case of the shipping industry, will comply with a trajectory towards zero emissions by 2050, so that financiers can make informed decisions about how their assets will contribute to climate action and mitigate the climate risks that could affect the valuations of their assets.[17]

An important characteristic of the shipping transportation industry is that a large part of the global fleet transports fossil fuels, such as primarily crude oil, coal and LNG-LPG. Thus, it would be counterintuitive to an extent to finance the building or even retrofitting of such vessels through a green bond issue. In addition, given the fact that the economic life of ocean-going commercial vessels exceeds 20 years and the objective has been set to transit into zero emission shipping up to 2050, CBI excludes explicitly crude oil tankers and LNG carriers from the CBI climate bond certification. However, the CBI criteria are technology-agnostic in terms of the energy required for the vessel to sail. Thus, any type of propulsion, including LNG propulsion, is eligible for certification provided it meets the emissions trajectory requirements.

3 Discussion and Conclusions

Despite the significant steps towards financing sustainable investments in the shipping markets, it seems that it is still early to evaluate their effectiveness. In fact, Nordhaus (2019) provides evidence that the overall anti-climate-change efforts are still insufficient to reduce projected emissions at the required level. Financial institutions and investors in general have the potential to contribute substantially in accelerating progress towards the goal of reducing the environmental footprint of the shipping industry. However, the following issues need to be resolved: (1) financing any green investments, shipping-related included, needs a definition and taxonomy of what is considered green and the degree of "greenness", an issue discussed in detail in Rebelo (2020), (2) national and/or voluntary initiatives, as opposed to global and mandatory ones, seem inadequate to deliver the set target (Nordhaus,

[17]A group of industry experts contributed to the development of the CBI shipping climate bond criteria, such as, the NYK, Nordea, Citi, ABN AMRO, etc.

2019), (3) commitments and targets set need not be modest but at least at par with the aspirational temperature targets (Nordhaus, 2019).

As discussed in this paper, taking action against climate change requires global cooperation which complicates the whole process. The shipping industry, being global in nature and a relatively highly emitting industry, is at the forefront of the global effort to reverse the effects of climate change. However, this goal is hard to achieve, despite the efforts of the IMO, as it requires the resolution of several conflicts of interests across several stakeholders involved, such as, the shipping market players and countries (flag states). Ship-lending capital providers are expected to have an important role to the whole process as described in this paper.

This study discussed the developing regulatory framework on green shipping finance and the role of capital providers in reducing the environmental footprint of the industry. Since 2018, there has been active involvement of national and supranational organizations, such as the IMO, the EU, etc., which put the shipping industry also on the forefront of reducing its environmental footprint at the global level. The ship financing market players can play an important role in the process and a number of initiatives on various fronts have been presented and discussed here. They include the Taskforce on Climate-related Financial Disclosures (TCFD) of the Financial Stability Board (FSB), the European Investment Bank (EIB), The Poseidon Principles, the Green Loan Principles (GLP), the Getting to Zero Coalition and the Climate Bonds Initiative (CBI). Combined, they exhibit the potential to help towards achieving IMO's set target for carbon emissions for the maritime industry.

Acknowledgements The authors would like to thank Ms. Stergiani (Stella) Moysiadou for excellent research assistance. Any remaining errors or omissions are the authors' responsibility.

References

Alexandridis, G., Kavussanos, M., Kim, C., Tsouknidis, D. A., & Visvikis, I. (2018). A survey of shipping finance research: Setting the future research agenda. *Transportation Research Part E: Logistics and Transportation Review, 115*, 164–212.

Basel IV. (2016) *BIS report.*

Garcia, B., Foerster, A., & Lin, J. (2020). *The shipping sector and GHG emissions: The initial strategy for a zero-carbon pathway. National University of Singapore Working Paper 2020/013.*

ICMA. (2017). *The green bond principles, International Capital Market Association.* https://www.icmagroup.org/Regulatory-Policy-and-Market-Practice/green-social-and-sustainability-bonds/green-bond-principles-gbp/

IMO. (2014). *The third IMO greenhouse gas study.* Published by the IMO.

IMO. (2018). *The IMO action to reduce greenhouse gas emissions from international shipping.* Published by the IMO.

Kavussanos, M. G., & Tsouknidis, D. A. (2014). The determinants of credit spreads changes in global shipping bonds. *Transportation Research Part E: Logistics and Transportation Review, 70*, 55–75. https://doi.org/10.1016/j.tre.2014.06.001

Kavussanos, M. G., & Tsouknidis, D. A. (2016). Default risk drivers in shipping bank loans. *Transportation Research Part E: Logistics Transportation Review, 94*, 71–94.

Kavussanos, M. G., & Visvikis, I. D. (2016). *The International handbook of shipping finance: Theory and practice*. Palgrave Macmillan Publishers.

Nordhaus, W. (2019). Climate change: The ultimate challenge for economics. *American Economic Review, 109*(6), 1991–2014.

OECD, et al. (2016). Green bonds: country experiences, barriers and options. http://unepinquiry.org/wp-content/uploads/2016/09/6_Green_Bonds_Country_Experiences_Barriers_and_Options.pdf

Rebelo, P. (2020). *Green finance for a sustainable maritime transport system: developing a universal vernacular for green shipping*. Working Paper.

Part III
Digitalization and Maritime Big Data

Maritime Digital Business Models: The Case of D/S Norden, 2015–2020

Martin Jes Iversen

Abstract This chapter concerns the digitalization of the maritime sector with a specific focus on business models. It is the argument of the article that current research in Maritime Informatics is focused on technological optimizations and thus lacks a commercial aspect in order to grasp the importance of digitalization in the shipping sector. In order to fill this gap a business model framework is suggested in the article with focus on the level of respectively customer-based-value-propositions and land versus sea. Then follows the empirical case of the Danish shipping company Norden and the development from 2015 to 2020. Norden is a leading commercial operator of dry bulk and product tanker vessels with more than 350 vessels in operation. The conclusion of the case is that Norden so far has regarded digitalization as tool for decision taking processes, which in the long-term should lead to compete advantages in terms of more efficient decisions based on big data and advanced algorithms. The shipping company has on the other hand decided not to use digitalization for the development of new software products and in accordance to presented digitalization matrix focused on indirect value proposition for the customers rather than direct customer-based initiatives. This focus confirms the hypothesis that digitalization in the dry bulk and tanker segment will often be based on indirect value propositions while digitalization in container-shipping might have a more direct relation to specific customer-based value propositions. This distinction is linked to the business-to-business nature of dry bulk and tanker and the more mixed business to business/business to consumer nature of container shipping—in. particular when the container shipping is integrated to the value chains and thus moved closer to the ultimate customers' preferences and services.

M. J. Iversen (✉)
Copenhagen Business School, Frederiksberg, Denmark
e-mail: mji.si@cbs.dk

B.-W. Ko, D.-W. Song (eds.), *New Maritime Business*, WMU Studies in Maritime Affairs 10, https://doi.org/10.1007/978-3-030-78957-2_7

1 Introduction

In the recent years international shipping has experienced major upheavals. Unstable markets and vague trade growth caused a trade volume expansion of only 0,5% in 2019 (UNCTAD, 2020). And the following year was marked by the intrusive Covid-19 pandemic. The "Great Lockdown" accelerated two mega-trends which already appeared before 2020. Firstly world trade was marked by shortened supply chains and further regionalized. This process took place due to an increasing tendency of home-shoring in terms of labour-extensive smart production. International maritime trade was increasingly marked by the predominance of Asian and Intra-Asian trade. In 2019 this one region constituted respectively 41% of the total goods loaded and 62% of the total goods unloaded (BCG, 2019; UNCTAD, 2020).

Secondly international shipping was affected by new consumer patterns in particular due to boosted e-commerce. According to UNCTAD, the estimated global sales value of e-commerce (including business to business and business to consumer) was equivalent to 30% of global GDP already in 2018 and the annual growth rates have exceeded 10% since (UNCTAD, 2020).

The legacy of the pandemic included a strengthening case for digitalization and dematerialization, as digitalization "... helped to maintain continuity in transport operations and trade processes ..." (UNCTAD, 2020, p. xiii). Policy actions were thus taken in order to prepare for a post-pandemic world marked by technology uptake and digitalization. The visions of a forthcoming digital transformation have concerned all major segments in shipping, but in various degrees and in different ways.

Container carriers find themselves in a race to build data analytics capabilities in an effort to pre-empt competition from giant e-commerce companies, such as, Alibaba and Amazon. Indeed, digital transformation has ben a hot topic in the container shipping industry. Consultants such as PwC, BCG and Deloitte predict a current need for new business models including trading platforms and more efficient institutional settings through blockchain technology (PwC, 2017; Vonck, 2017; BCG, 2019).

Bulk shipping operators have begun developing digital platforms for streamlining vessel operations and commercial shipping. Norwegian-based Torvald Klaveness, for example, has been fast to develop and market digital platforms for monitoring, benchmarking and valuing vessel performance. Western Bulk, another Norwegian-based bulk operator, is building a digital platform for improving internal company communication and enable flexible working practices.

In the tanker segment commercial decision-taking systems have been initiated on the basis of big data analyses. More important the "integrated ship" idea is now taking shape with a constant flow of information linking the vessels with the corporate on-shore facilities in order to optimize trading routes and ensure the technical maintenance. This last feature runs across the segments with a tendency to be most important in the capital-intensive segments such as LNG, ultra large container vessels and offshore.

From a broader stakeholder perspective, national governments and supra-national bodies push for policy and investments to advance digital transformation in shipping, ports and maritime industries (European Commission, 2016; Norway—Ministry of Trade, Industry and Fisheries, 2016; Denmark—Ministry of Industry, Business and Financial Affairs, 2018), international consultancy companies make a living from pointing out the opportunities and barriers to digital transformation in the maritime sector (Futurenautics Maritime in association with Ericsson, 2017; pwc, 2017; Saxon and Stone, 2017a, 2017b; Vonck, 2017; Quitsau et al. 2018; Egloff et al., 2018), and there is considerable caveat arising from maritime media and interest organizations that we must act now in order not to be left behind (Nightingale, 2017; Lloyd's List, 2018). Indeed, according to Danish Maritime, an industry association catering to the interests of Danish yards and maritime equipment suppliers, first-movers are needed now and for the future (Danish Maritime, 2015).

The experiences of the industry before and during Covid-19 and recent initiatives by the authorities leaves little doubt that digitalization will become increasingly important for international shipping in the future. The question is how to analyse and understand the phenomenon of maritime digitalization. In a recent book entitled "Maritime Informatics" the focus is on the collection and recording of information that supports maritime operations and trade (Ward & Bjørn-Andersen, 2021). The purpose of gathering digital information is to enhance knowledge and thus support decision, which "enhance a shipping company's capacity to compete with others in the maritime sector" (Ward & Bjørn-Andersen). Despite this commercial ambition the focus of the maritime informatics field (and thus the book) is primarily technological including the need of standards and social-technical systems which enable information gathering to support decision-taking processes for instance in relation to voyaging, port visits and general supply chains.

The argument of this article is that the Maritime Informatics' focus on collecting and recording information that supports maritime operations and trade lacks the important commercial dimension of digitalization. The commercial dimension must include customer relations and this particular aspect of maritime digitalization is absent in the conventional definition of maritime informatics. This chapter aims to bridge the peculiar empirical experiences of a specific shipping company with the specific theoretic customer focus of the business model approach. Three sections follow in the chapter: firstly an introduction to the theoretic concept of business models including an attempt to bridge the scholarly definition with the industrial logics of the shipping sector. Then follows the case of the digitalization in the Danish shipping company Norden including an analysis of the relevant business models and finally the conclusion.

2 A Business Model Framework for Maritime Digitalization

Scholars of strategy and management use the notion of a business model to describe a firm's key processes and how they are related (Zott et al., 2011). A business model can be defined as the "design or architecture of the value creation, delivery, and capture mechanisms" of a company to (Teece, 2010). Business model innovation for digitalization requires companies to reconfigure their value proposition (what is being offered to customers) as well as the way in which that value is created and delivered (value delivery). Companies have tended to focus on mainly one of these areas through a set of particular initiatives (Berman, 2016). In general, the more components of a company's value creation and delivery system that are simultaneously changed during such initiatives, the more complex the form of business model innovation (Taran et al., 2015). Rather than simply categorizing digital transformation projects in shipping along the dimensions of business model complexity and customer benefits, it is useful to look instead at the process through which business models emerge in a company). It has suggested that the core logic of a business model revolves around the firm's value proposition to the customers with the important point that business models ". . . can be a vehicle *for* innovation as well as a subject *of* innovation." (Zott et al., 2011). Such a discovery- and innovation-driven approach to digital transformation provides a suitable way of framing the challenge that businesses face when confronted with uncertainty, complexity and fast-moving and unpredictable business environments such as the global shipping market in 2020. The discovery-driven approach was first introduced in the strategic planning literature (McGrath and MacMillan, 1995) as an approach for companies to minimize expenses and maximize learning in their planning efforts, changing the focus of strategic change away from one of planning based on analysis and projections towards one based on experimentation and learning, prying managers to consider managing expenditures with discipline, being conscious about key assumptions made regarding the decision-situation at hand, and exhausting all possible ways to create new knowledge before making irreversible commitments.

Figure 1 provides a matrix to position digital transformation initiatives in shipping on the dimensions of the business model focus on customer-based value-proposition combined with the technology focus on on-shore versus on-ship. While a full analysis of the business model implications of digital transformation in shipping would be beyond the scope of this article, the matrix is indicative of the current digitalization initiatives in the maritime sector. A broader analysis would necessarily take into consideration the many differences in company organization and market dynamics pertaining to different industries.

The matrix of digitalization in shipping combines two dimensions. Firstly the y-axis concerns the degree to which the maritime digitalization is based on respective direct and indirect customer based value-proposition (Chesbrough & Rosenbloom, 2002). An example of digitalization based on a direct customer-based value proposition could be the booking platform Maersk Spot introduced in

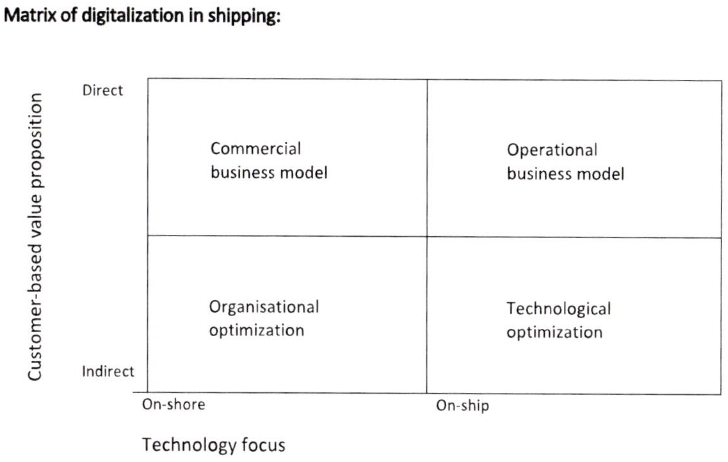

Fig. 1 Matrix of digitalization in shipping

June 2019, which is based on four principles: easy online booking, fixed price at booking, guaranteed loading and finally inland and ocean in one booking (https://www.maersk.com/transportation-services/maersk-spot). In contrast a digitalization initiative based on an indirect customer-based value proposition is defined as *optimization* rather than a *business model*. Such as an optimization could have a benefit for the customers but it would be indirect and thus not been designed with the specific purpose in mind. An example of an indirect customer based-value proposition could be the s-Suite from StormGeo which combine weather-based voyage planning with onboard route advisory service and fleet performance management (https://www.stormgeo.com/products/s-suite/).

The second component of the matrix—illustrated by the x-axis in the matrix above - concerns the technology focus which in shipping can be divided between on-shore versus on-ship. The distinction between on-shore and on-ship is peculiar and critical for shipping. One of the most important challenges in shipping has historically been to integrate the (commercial) decision on shore with (technological) decisions on the vessels. In many cases this has led to wide gap between stakeholders based on shore and on sea—including separate companies for the commercial landbased operations and the technical sea-based operations of the vessels. Since the 1990s this distinction has been enforced by the outsourcing of specific functions to specialized ship-operation companies. This specialization—with separate companies designed a specialized for specific purposes—will probably continue, but it is important to keep in mind that digitalization offers a new and increasingly strong opportunity for integrating land- and sea-based activities. A good example is the opportunity via censors to control and monitor both the routeplanning and the technical maintenance of the vessel, and on the other hand the opportunity for the seafarers to follow and discuss technical, organisational and commercial matters on shore via internet-meetings and conferences.

The matrix provides four variations of digitalization in shipping:

- The top-left combination of high focus on customer-based value proposition and on-shore technology focus which can be defined as commercial business model. This could for instance be a trading platform in liner-shipping;
- The low-left combination a low customer-based value proposition and a on-shore technology focus. An example of such "organizational optimization" could for instance be internal digitalization of the HR and recruitment systems;
- The top-right combination of high customer-based value proposition and on-ship focus is defined as an operational business model, and this could for instance be real-time updates on cargo movements;
- Finally low-right combination of a low customer-based value proposition and the technological focus on ship is labeled technological optimization and an example could for instance be the introduction of new digital navigation systems, a satellite based weather tracking system or digital engine maintenance systems.

It is important to emphasize that the matrix illustrates "archetypes" of digitalization. Many empirical cases will concern hybrids with aspects from several categories. It is never-the-less important to define which kind of digitalization that are relevant for the analysis as digitalization without definitions becomes a blurred and unclear phenomenon. The hypothesis is that most digitalization initiatives will cluster in two categories of the matrix: the top-left, "commercial business model" category and the bottom-right "technological optimization" category as on-shore initiatives tends to have a commercial performance-driven dimension while the on-ship digitalization initiatives tends to have a technology focus in terms of operational optimization of the vessels.

3 The Case of Digital Transformation in D/S Norden, 2015–2020

This section of the article concerns the development of the Danish shipping company Norden from 2015 to 2020 with a specific focus on the digitalization strategies (Iversen, 2021). Norden was founded in 1871 and in 2015 the shipping company operated a fleet of app. 180 dry bulk vessels and 80 product tankers. Following a very difficult time period after the financial crisis in 2008 the top management of the company was changed in April 2015 when Jan Rindbo became CEO. During the first 100 days this CEO went all the way around Norden. Departments, ships, branches, customers and contacts were visited, reviewed and understood. For the new CEO, it was all about listening, learning and absorbing. He wanted to get an understanding of the Company's strengths and weaknesses, challenges and opportunities.

The results of Rindbo's 'tour des departments' were presented at the Board Meeting of 11 August 2015. Rindbo said that he had understood four characteristics of Norden. First, the Company was characterised by a competent and dedicated staff.

To this one could add a very strong reputation in the industry, where Rindbo noted how 'all doors were open.' The last two points were challenging. First, the new CEO had noted a highly complex organisational structure. When he reviewed the financial analysis, it was difficult to pinpoint where money had been lost or earned. Which departments, decisions and ultimately staff really contributed to the Company's profits and losses? It was difficult to identify. The fourth and last point concerned the core fleet, which Rindbo had noted was spread over eight ship types—five in dry cargo and three in tankers.

Under the heading 'Focus & simplicity', Rindbo's strategic answer to Norden's challenges consisted of three independent facets. First, Norden had to aim for fewer, smaller, and more flexible ship types in future. At the end of 2015, Norden operated 184 dry cargo ships: 77 Supramax ships (50–64,000 DWT), 55 Panamax ships (75–93,000 DWT), 40 Handysize (28–39,000 DWT), nine Post-Panamax (110–120,000 DWT) and three Capesize ships (170–180,000 DWT). The focus would now be on the Panamax, Supramax and Handysize ships, where the Company had the strongest market position and had the opportunity as an operator to create extra value beyond what was generated by the long-term market direction (Iversen, 2021).

The concentration on the three most popular ship types was related to the second facet of the strategic plan, which was a more adroit approach to the markets as an operator in the short-term market combined with opportunistic purchases and sales of ships in the long-term market in order to constantly adapt the fleet portfolio to market conditions.

The third element in the strategy was the building of a 'performance culture' consisting of two basic elements: empowerment of the staff through decentralisation of responsibility, and full transparency in the creation of profits. Key Performance Indicators (KPIs) were to offer an insight into the contributions of small groups (teams). Decisions were to be made with the aid of new data-based decision-making processes. Digitalisation was to top Norden's agenda (Iversen, 2021).

After the presentation came the implementation. Rindbo had 15 years of work experience in Hong Kong and thus a thorough knowledge of Asia. After the Board Meeting on 11 August 2015, a noteworthy analysis of the Chinese market landed as a memorandum on the CEO's desk (Buhl, 2020). The key to understanding Norden's development in the crucial years from 2000 to 2015 lay, according to the memorandum, in the development of demand in the huge Asiatic country. From 2001 to 2014, China's steel production had risen by 540%, while the related Chinese imports of iron ore rose by 840% (Buhl, 2020). It was these demand figures that had driven the dry cargo rates up into the skies from 2004 until 2008. The point of the memorandum was that this was an exceptional period, and that it was *not* likely that a similar growth in demand would be repeated. The roads had been built, the high-rises erected, and the bridges constructed. Large amounts of coal, bauxite, iron ore and grain would still be imported. But whereas the growth rates of the 'golden age' in Chinese raw material imports were 12–15% a year, the market was now approaching a more normal annual growth rate of around 5%. The figures reflected the fact that the annual Chinese growth in GNP had dropped from around 10% in the period

2000–2010 to around 5–7%. On the other hand, the total global order book for new dry cargo ships for delivery in 2016 represented some 10% of the world fleet. The reality was, therefore, that despite a weakening market, approximately 45 million DWT in new dry cargo ships would be delivered in 2016—the same high level as in 2014 and 2015 (Norden, 2014, 2015). The recipe thus lay on Jan Rindbo's desk. Hitherto, the pattern had been that when dry cargo rates rose, Norden earned a lot of money, but when the rates dropped, the Company's profits took a dive. If the market's low level continued, Norden would follow suit in the form of large billion-kroner write-downs that eroded equity capital—a development that could theoretically threaten the Company's existence. The aim for Rindbo now became to change this course. *In the future, Norden should be able to earn money in any dry cargo market.* This was the radical goal that required two profound strategic and organisational changes. Rindbo realised that the whole organisation had so far worked on the basis of a framework laid down by the Management in the corner office. If it was decided to contract 20 large ships, this one decision laid the framework for the subsequent results—both good and bad. Instead of one 'CEO's bottom line', the new CEO wanted many smaller bottom lines. Norden had to be able to manoeuvre quickly and flexibly when the unpredictable market went in different directions—both upwards and downwards. The only way to do this was to get the staff's independent decisions into play. The grand gold-plated 3-year strategies were passé. Now was the time for many shorter-term decisions. But this required transparency, competence, and manoeuvrability (Iversen, 2021).

From 2014 until 2016, the fleet of company-owned dry cargo ships was therefore reduced from 26 to 18. In the same period, no new ships were contracted. The slimming-down of Norden's core fleet was followed in July 2017 by the division of the Dry Cargo Division into two independent units: Dry Cargo Operator and Dry Cargo Owner. In the former, earnings were to be generated in three ways n (Norden, 2018):

- Matching cargo and ships in a logistical and operational optimisation
- Exploitation of the Company's size, market knowledge and access to clients and ships
- Shorter market positions

As early as August 2015, Jan Rindbo proposed that Norden should supplement the operating activities with 'asset play', an opportunistic approach to buying and selling ships. However, it was important for the new Management that the earnings and losses from ship transactions were separated from the operating and chartering earnings. Everyone should be able to see where the money was lost or earned, and each department had to be able to work within market conditions—for example, whether to charter a Norden ship out to other customers, or involving chartered-in ships, if it made commercial sense. The new Dry Cargo Owner department therefore had to create long-term results with the aid of three methods:

- Planning and negotiating favourable agreements on owned ships as well as long-term chartering.

- Making use of Norden's long-term relations with yards and shipowners with possible access to 'off-market' ship deals.
- Creating a portfolio of long-term—and thus safe—contracts of affreightment.

Norden's new organisation made sense. From the summer of 2017 until the end of 2019, there was a surplus in the Dry Cargo Operator department every year. In that period, a total surplus of DKK 333 million was generated without taking great risks on the market in relation to the traditional operations of the Company. The goal was to be able to make a profit in any market, and the means were via flexible teams, supported by digital technology. Since Norden had operated hundreds of different ships over many years, a huge amount of data had gradually been gathered—on among other things fuel consumption, weather routing, cargo conditions, speeds and so on. The challenge now consisted of creating value from that data. Norden could thus gain an advantage by being a big player in a fragmented market where small new competitors with low cost bases regularly entered the market. A new department was established with the long title 'Advanced Analytics and Digital Solutions.' The staff of 11 were trained engineers, computer scientists and quantitative analysts, and their roles included developing and optimising algorithms that enabled every employee in Norden to charter the right ship, at the right price, in the right place. Many factors had to be considered, including the right speed, the cargo type, the weather, and the route (Iversen, 2021).

Basically, the digitalisation at Norden was divided into three areas. First, the digital solutions were used for process optimisation. This was a matter of optimising order influx and data flow in the daily activities of the organisation. This part of the work did not differ much from any other organisation—within and outside shipping. The second area was 'analytics and decision support.' The decision in Norden was that the analytical 'big data' competencies were to be built up internally in the organisation. In this way, new powerful tools were gradually created, which the Company's charterers and other core employees could use when commercial decisions were to be made; tools, that is, to which the competition had no access. The whole idea was to build up competitive advantages by being a major player in a fragmented market.

The third and last area was digital products. This could for example be the building-up of a platform for booking cargo or the sale of analytics and decision-making software to other shipping companies and market players. Norden decided against taking this route. It was assessed that the disadvantage of selling solutions and data to competitors outweighed the advantages of gradually also becoming a software company.

The winter of 2019 became the department's big test. At that time, rates had halved within a few weeks after the collapse of a major dam at Vale, the major mining company in Brazil, which cost many people their lives. The event led to a decline in the transport of iron ore between South America and China. However, the Company did get through the quarter with a small profit, and subsequently Jan Rindbo reiterated the advantages of having two income sources in an unpredictable business: 'Norden generated a positive result in a quarter characterized by a strong winter market in Tankers and a Dry Cargo market on the brink of collapse.... not even a fast and adroit reaction by the Dry Cargo Operator could compensate for a halving of rates within a period of three weeks. Dry Cargo Owner, however, was well protected against the weak market thanks to a high level of coverage and generated a break-even result; Tanker Operations were well positioned in a very strong winter market and achieved a good result.' (Pico, 2019).

At the beginning of 2020, the 'Operator' structure used by the Dry Cargo Division was also introduced at Norden's Tanker Division. Norden now consisted of three business units. First, Dry Cargo Owner changed to Asset Management. This new division combined tankers and dry cargo in terms of purchases, sales, and long-term charters in respect of the core fleet, as well as the technical operation of owned ships. Secondly, there was still Dry Cargo Operator, and finally Tanker Operator, where this newly created department was built on the following three principles (Iversen, 2021):

- Maximising the operational margin between spot earnings and time-charter costs by optimising route planning, choice of ships and capacity utilisation
- Maximising the advantages of geographical and segment-based positioning on the basis of digital analyses as well as market expertise
- Commercial operation of third-party tankers

The last item was important and made Tankers different from Dry Cargo. At the end of 2018, Norden took over 100% of the ownership of Norient Product Pool ApS, which at a stroke led to a doubling of the Tanker Division's earning potential. Currently, Norden gained 64 'new' full-time staff. Norden owned 24 product tankers at the end of 2019, while Norient operated a total of 107 ships. The number increased by 28 MR tankers in June 2020, when the American based Diamond S decided to place their ships in the Norient Pool (Carlsen, 2020). With this addition, Norient comprised 139 ships—and as such, in first place among the world's largest product tanker pools in the Handysize and MR segments. It was noteworthy that four of the world's five largest product tanker pools, as of the summer of 2020, were based in Copenhagen—and thereto was Torm, which whilst no longer operation pools, was one of the biggest players globally in this segment.

Diversity has several dimensions, and in 2019 there were 32 different nationalities employed in the Danish Company, where English had become the working language. On large monitors in Hellerup there was constant interactive contact with colleagues in Asia, Australia, Africa, South and North America. The point of having 10 Norden offices on six continents was to be close to the customers. But it was the digital technologies that made possible the close and flexible relationship between the departments and the head office. In the spring of 2020, the digital relations took on an unexpected and sudden leading role. In a few weeks, the world was struck by a fearsome and infectious corona virus, COVID-19, and like other Danish companies Norden closed its office facilities. But the Company carried on. Meetings were held, plans developed and executed virtually. As an organisation, Norden was prepared for the changes—the staff were accustomed to meeting virtually—and whether a colleague sat in Ballerup or Bombay turned out not to be of such great significance.

When the offices closed, the ships continued to sail. The raw materials for the world's industries and consumers still had to be delivered. Jan Rindbo described the situation in a posting on LinkedIn in mid-April 2020, when societies on land were paralysed by lockdown: 'Just now, the fleet is particularly busy delivering indispensable grain and energy products to keep the wheels rolling in world trade. It is important that ports continue to function, thus keeping the global supply chains secure.' (Rindbo, 2020). For the seafarers, the COVID-19 crisis presented the special challenge in that it became difficult to implement crew changes, meaning that many seafarers could not leave their ships at the end of their contracts—a critical situation that required political action and international solutions. Norden therefore participated in a global COVID-19 alliance organised by the Singapore-based management company Synergy Group, representing 1500 ships and 70,000 seamen, to which one could add support from both the International Chamber of Shipping (ICS) and the International Transport Workers' Federation (ITF) (Hartkopf-Mikkelsen, 2020).

The idea of quantifying and establishing guidelines for a company's social responsibility (CSR) caught on in Denmark after the year 2000. In 2009, Norden signed the UN's Global Compact, and over the years the Company's CSR goals took form. A quantum leap was taken in 2019, when a CSR strategy was launched with a special focus on the UN's Sustainable Development Goals (SDGs). The goals were originally adopted in 2015 when the Danish politician Mogens Lykketoft was Chairman of the UN General Assembly in New York. SDG number nine, 'Industry, Innovation and Infrastructure', was identified in 2019 as the pivotal one for Norden, with a special focus on four areas (Iversen, 2021):

- Climate and environment—CO_2 emissions, spillages, SOx, NOx and plastic;
- Human beings—safety, human rights, and employees;
- Anti-corruption—compliance, education and a whistleblower scheme;
- Responsible partnerships—primarily with selected NGOs, among other things in relation to the combating of plastic pollution.

A new and goal-based job function with responsibility for accelerating the development of CO_2-neutral transport was established in September 2019. During the decade from 2007 to 2017, Norden had reduced CO_2 emissions per cargo ton transported on its own tankers by 25%. 'We will further strengthen our efforts,' stated Jan Rindbo in the autumn of 2019, 'to help guide the shipping industry towards a cleaner future and offer our customers CO_2-neutral transport. We have come a long way by increasing our fuel efficiency, but there is a need for initiatives, for example the testing of new CO_2-neutral fuels, to reach the targets.' (Norden, 2019). At the end of September 2018, the 37,000 DWT Handysize product tanker *Norden Highlander* sailed from Rotterdam in the Netherlands to Tallinn in Estonia on the first voyage to test a new $C0_2$-neutral biofuel (Norden, 2018). The collaboration with the organisation GoodFuels continued, and another test voyage was performed in 2019. The first tentative steps on a long, challenging, and ambitious voyage into the future had been taken by the 150th anniversary. The goal is a 40% reduction of the ships' transport emissions by 2030, to be followed by a 100% CO_2-neutral Norden fleet in 2050 at the latest.

4 Conclusion

There are at least three important lessons to learn from the digitalization strategies at Norden from 2015 to 2020. Firstly digitalization proved pivotal to the Rindbos new "Focus & Simplicity" strategy. The aim was *commercial,* but not as directly customer-based value proposition. Rather the aim was to optimize the internal operations on shore through the creation of decentralized business units, and in this way make it transparent where money was lost or earned. The precondition for this agility was digital platforms, which played two roles: firstly the internal procedures and operations were optimized through digitalization for instance of internal financial and legal transactions and secondly digital tools were created in terms of databases, AI tools and platforms. In the new department "Advanced Analytics and Digital Solutions" coordinated and executed the digitalization. The initiatives were divided in three categories:

- digital solutions for optimization of processes,
- digital decision-support,
- finally services (platforms) for customers.

The emphasis in Norden had so far been on the first and second component in order to achieve competitive advantages being a large player in fragmented markets.

The second important lesson to learn is connected to the three categories mentioned above. The digitalization process in Norden was commercial but at the same time focused on initiatives that were indirect in the relation to customers. In the matrix below Nordens digitalization initiatives should be placed in the lower parts of the matrix. The initial restructuring of the company focused on organisational optimization while the tools for decision taking could be defined as a combination of organizational and technological optimization. The next step for Norden—in terms of digitalization—could be step upwards towards more direct customer-based value-propositions. These could be customer platforms and options easy booking as known from liner-shipping for example Maersk Spot mentioned above. A suggestion for further research could be to investigate how and why digitalization initiatives mirrors the various shipping segments, and not least how for instance dry bulk shipping with an apparent concentration in the lower end of the matrix could learn from digitalization experiences in the liner-segment, which seems to be placed higher in the matrix.

Matrix of digitalization in shipping

	Direct	Commercial business model	Operational business model
Customer-based value proposition	Indirect	Organisational optimization	Technological optimization
		On-shore	On-ship
		Technology focus	

The third important lesson to learn from the Norden case is that the digitalization effort have accelerated, but not changed course, during the Covid-19 crisis in 2020. During the "Great Lockdown" Norden, along with other shipping companies, was able to continue its operations including the transportation of life stocks and coal thanks to the existing digital infrastructure. Employees from the headquarters and the branches could continue their functions more or less independent of the physical origin of the offices—or in this case home offices. At the same time Norden did not use the Covid-19 crisis to expand its digitalization in the direction of a more direct customer-based value proposition. It will be interesting to follow digitalization in dry bulk companies such as Norden in the coming years and thus see to which extend companies in this segment will integrate the organizational and technological optimization with commercial and operational business models.

References

BCG. (2019). *Digital Transformation in the Shipping Industry*. Available from https://www.bcg.com/industries/transportation-travel-tourism/center-digital-transportation/shipping.aspx

Berman, S. J. (2016). Digital transformation: Opportunities to create new business models. *Strategy & Leadership, 40*(2), 16–24.

Buhl, J. (2020). Norden: The Critical China Note' in Iversen and Buhl (2020). In *Danish Shipping in the 21st Century* (pp. 41–51). Palgrave MacMillan.

Carlsen, C. (2020). Norden slog flere konkurrenter i kampen om Diamond S-skibe, *ShippingWatch*.

Chesbrough, H., & Rosenbloom, R. S. (2002). The role of the business model in capturing value from innovation: Evidence from Xerox Corporation's technology spinoff companies. *Industrial and Corporate Change, 11*, 533–534.

Danish Maritime. (2015). *Danish maritime: Priorities must be given to first movers*. Available from https://splash247.com/danish-maritime-priorities-must-be-given-to-first-movers/

Egloff, C., Sanders, U., Riedl, J., Mohottala, S., & Georgaki, K. (2018). *The Digital Imperative in Container Shipping*. Available from www.bcg.com/publications/2018/digital-imperative-container-shipping.aspx

European Commission. (2016). *Commission staff working document on the implementation of the EU Maritime Transport Strategy 2009–2018*. Brussels: SWD(2016) 326 Final. Available from https://ec.europa.eu/transport/sites/transport/files/swd2016_326.pdf

Futurenautics Maritime in association with Ericsson. (2017). *Waypoint Digital 2017. Digital infrastructure, investment, initiatives and transformation in the shipping and maritime industry*. Futurenautics Ltd..

Hartkopf-Mikkelsen, J. (2020). 'Norden støtter internationalt opråb om besætningsskifte.' *Søfart*, 15 April.

Iversen, M. J. (2021). *The Norden Voyage*. Lindhardt & Ringhof.

Lloyd's List. (2018). *Maritime can't resist digital disruption for much longer*. A Roundtable discussion hosted by Lloyd's List, July 3, 2018. Available from https://lloydslist.maritimeintelligence.informa.com/LL1123257/Maritime-cant-resist-digital-disruption-for-much-longer

McGrath, R. G., & MacMillan, I. C. (1995). Discovery driven planning. *Harvard Business Review, 73*(4), 44–54.

Ministry of Industry, Business and Financial Affairs. (2018). *Maritime Denmark. A global maritime power hub*. The Danish Government.

Ministry of Trade, Industry and Fisheries. (2016). *Maritim21 – An integrated maritime strategy for research, development and innovation*. The Research Council of Norway.
Nightingale, L. (2017). Wilhelmsen chief calls for Silicon Valley mindset. *Lloyd's List*, May 30, 2017.
Norden. (2014). Annual Report.
Norden. (2015). Annual Report.
Norden. (2018). Annual Report.
Norden. (2019). Annual Report.
Pico, S. (2019). 'Norden trukket ned af 'tørlastmarked på kanten af kollaps.' *ShippingWatch*.
pwc. (2017). *The digital transformation of shipping. Opportunities and challenges for Norwegian and Greek shipping companies*. Available from www.pwc.no/no/publikasjoner/shipping/The-Digital-Transformation-of-Shipping_HE-NO.pdf
Quitsau, J., Tontara, N., Vöpel, H., Jahn, M., Otto, A., & Wolf, A. (2018). *Shipping in an era of digital transformation*. Berenberg and HWWI.
Rindbo, Jan: Linkedin, 2020.
Saxon, S., & Stone, M. (2017a). *Container shipping: The next 50 years*. McKinsey & Company – Travel, Transport & Logistics. Available from www.mckinsey.com/~/media/mckinsey/industries/travel%20transport%20and%20logistics/our%20insights/how%20container%20shipping%20could%20reinvent%20itself%20for%20the%20digital%20age/container-shipping-the-next-50-years-final.ashx
Saxon, S., & Stone, M. (2017b). *How container shipping could reinvent itself for the digital age*. McKinsey & Company – Travel, Transport & Logistics. Available from www.mckinsey.com/industries/travel-transport-and-logistics/our-insights/how-container-shipping-could-reinvent-itself-for-the-digital-age
Taran, Y., Boer, H., & Lindgren, P. (2015). A business model innovation typology. *Decision Sciences, 46*(2), 301–331.
Teece, D. J. (2010). Business models, business strategy and innovation. *Long Range Planning, 43* (2/3), 172–194.
UNCTAD. (2020). *Review of maritime transportation*.
Vonck, I. (2017). The impact of maritime digital transformations. Three options to respond to the challenges of digital disruption. Deloitte, August 8. Available from www2.deloitte.com/nl/nl/pages/energy-and-resources/articles/the-impact-of-maritime-digital-transformations.html
Ward, R., & Bjørn-Andersen, N. (2021). The origins of maritime informatics. *Maritime Informatics, Springer Press.*, 5–11.
Zott, C., Amit, R., & Massa, L. (2011). The business model: Recent developments and future research. *Journal of Management, 37*(4), 1019–1042.

Platform Revolution in Container Shipping Markets: An Economics Perspective

Byoung-Wook Ko

Abstract This article investigates so-called platform revolution in container shipping markets from a perspective from economics. Drawing on a huge economic literature on platform expansions, the author argues that there is and will be enormous economic benefits from platform revolution in general and in container shipping markets. For the exposition for this argument, this chapter, first, reviews the literature for the conceptualization of platform phenomena, especially focusing on the network economics. Second, it suggests the two interesting and important examples of platforms in the global container shipping markets. One is the blockchain platforms such as TradeLens and GSBN. The other is the platform-related market price index, i.e., FBX. Then, this study looks into the prospective benefits of, barriers to, and threats from the platform revolutions in container shipping markets. Finally, this article concludes by suggesting that the market players should act as soon as possible for the adaptation to this rapid digitalization including platform revolution in order to not to exit from market but to develop and prosper in the future.

1 Introduction

Figure 1 shows the top five companies in U.S.A. by stock prices. These companies are common in the sense that they are all global leader digital platform companies. For example, Apple provides the smartphone application platforms and the famous google-map is one application. Amazon provides a comprehensive e-commerce online platform. Likewise, MicroSoft, Google, and Facebook also provide the huge platforms to our society.

In addition, Table 1 shows other samples of platform companies. For example, consider the case of YouTube. This digital-intensive platform is making the entry and exit costs very lower, compared with the traditional service providers. Via

B.-W. Ko (✉)
Korea Maritime Institute, Busan, Republic of Korea
e-mail: valiance@kmi.re.kr

B.-W. Ko, D.-W. Song (eds.), *New Maritime Business*, WMU Studies in Maritime Affairs 10, https://doi.org/10.1007/978-3-030-78957-2_8

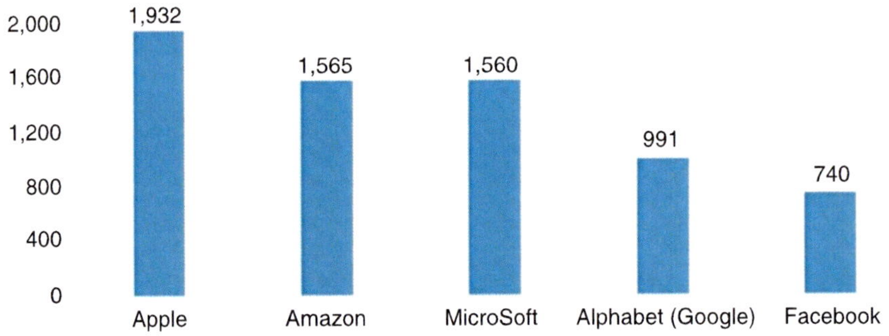

Fig. 1 Top Five Companies in U.S.A. by Stock Prices (3, Oct. 2020). Unit: billion $. Source: Google

Table 1 Examples of platform companies

Area	Company	Area	Company
Agriculture	John Deere	Labor	Upwork
Communication	Facebook	Local Services	Yelp
Consumer Foods	Philips	Logistics	Munchery
Education	Udemy	Media	YouTube
Energy and Heavy Industry	Nest	Operating Systems	Android
Finance	Bitcoin	Retail	Amazon
Health Care	Cohealo	Transportation	Uber
Gaming	Nintendo	Travel	Airbnb

Source: Parker et al. (2016)

YouTube, the providers of video contents can simply upload their videos and also the consumers of these contents can simply watch these videos. Imagine the traditional TV services. The entry barrier to video services had been very large. Further, the TV viewers had very small options for watching video contents. In short, the famous platform providers are making new markets with very low transaction costs.[1]

Following these digitalization trends, the global container shipping sectors are now transforming themselves into the digital-intensive organizations. The representative examples are the blockchain platforms across the supply chains such as TradeLens and Global Shipping Business Network (hereafter, GSBN). Also, in 2019, Maersk's share of on-line booking volumes is 25% (about 6.6 million TEUs) and Hapag-Lloyd's share is 15% (about 1.6 million TEUs). These mean that the various methods using digital platforms in container shipping markets are now enabling and facilitating more and more process and product innovations. Also, since 2020, the Freightos Baltic Global Container Index (hereafter, FBX) of platform character has been published on daily basis. So, a thorough review of relevant

[1]Transaction costs are the total costs of making a transaction, including the cost of planning, deciding, changing plans, resolving disputes, and after-sales (Wikipedia).

concepts, recent trends, and prospects (which is positive or negative) of platform revolutions in container shipping markets is worthy of attention, research, and reading.

This chapter consists of the five sections including this introduction. Section 2 explains briefly the concepts about platforms in economics literature and the following Sect. 3 helps for the readers to understand the platform structures in container shipping markets such as the blockchain platforms for supply chains and FBX for market indicators. Section 4 shows the prospective benefits, barriers, and threats of platform revolutions in container shipping markets. Finally, Sect. 5 concludes.

2 Platform in Concept

For the explanation of the concepts of platforms, I intensively draw on the economics literature.

2.1 *Definition of Platform*

First, it is necessary to make some distinctions among the related terminologies such as network, intermediary, market, and platform. For example, the globally famous book, "Platform Revolution" by Parker and his co-authors has the sub-title, "How networked markets are transforming the economy and how to make them work for you" (Parker et al., 2016). However, before going to the interrelated terminologies, I should say my definition for the platform. I define that 'platform is a form of intermediary using new digital technologies with the features of network and market'.

Then, I suggest the definitions of the related concepts. An intermediary is "a third party that offers intermediation services between two parties" (Wikipedia). A (social) network is "a social structure made up of a set of social actors" (Wikipedia). A market is "one of a composition of systems, institutions, procedures, social relations or infrastructures whereby parties engage in exchanges" (Wikipedia).

However, platforms are not a new phenomenon. For example, the traditional newspapers, academic journals, and fairs (i.e., markets) all belong to the cases of platform. However, the remarkable development of ICT (Information and Communication Technologies) has allowed us to scale up these platforms. On digital platforms, we do not need to appear physically in some places, and also copies can be made for free.[2] As I mentioned before, these conveniences mean the huge potential of virtual exchanges among platform participants. Facing a large variety of platforms, the analysts including economists have been trying to define the core

[2]For more detailed explanation in this aspect, refer to Sanchez-Cartas and Leon (2019).

functions of platforms. Relating to these efforts, the two-sided market theory is very useful for understanding new digital platforms.

Rysman (2009) defines that "a two-sided market is one in which two sets of agents interact through an intermediary or platform, and the decisions of each set of agents affect the outcomes of the other set of agents, typically through an externality". He said that the literature on two-sided markets (or platforms) could be seen as a subset of the literature on network effects. So, for more understanding the characteristics of platforms such as the aforementioned externalities, we need to review on the network effects.

2.2 Network Effects

The past accumulated researches classify the network effects into (1) the direct network effect and (2) the indirect network effect. In a huge literature on these direct and indirect effects, I would like to introduce the explanation of Gawer (2014), which is as follows: The direct network effect can be called as same-side network effect. This means that the benefit to a platform user depends positively on the number of other users in the same-side. The indirect network effect can also be called as cross-group network effect. This means that the benefit to a platform user depends positively on the number of other users in the other group. As will be explained later, the benefits of the blockchain platforms in container shipping markets such as TradeLens and GSBN, rely on the indirect network effect (i.e., cross-group network effect). In contrast, the benefits of FBX rely on the direct network effect (i.e., same-side network effect). The reason is that more logistics firms participate, better FBX's indicator represents the current market situation.

By the way, these network effects make so-called 'chicken-and-egg problem'. As a typical example, consider the following joke: "Without the telephone user, it is not necessary to use telephone. So, the prize for greatest salesperson in history should go to whoever sold the first telephone" (Parker et al., 2016). In addition, it should be noted that these positive network effects allow the 'winner take all markets' through the spiral feedback mechanism. This means that the leader platforms are more likely to have larger market share than the follower platforms. The mechanism of this 'winner take all markets' is that, as more users participate in a platform, the platform has more competitive advantages in the aspects on its use values, cost structures and so forth. For example, consider a user who can participate in a global large platform and a local small platform. If other things equal, this user will benefit more from the global platform than from the local one. In addition, as a platform has more and more participants, it can remarkably reduce the relevant average and marginal costs. This cost advantage will also enhance its competitiveness. This spiral feedback mechanism will allow the first-movers to exploit the potentials of the related markets and platform fields such as e-commerce markets, online meetings in the untact-and-ontact era of COVID-19 crisis, application apps markets, etc.

2.3 Four Lens for Platform

According to my literature study, Kiesling (2018) might be the best article for showing the diverse aspects of platforms. Adopting the Gawer (2014)'s taxonomy of technology-economic-organizational platform and adding her own epistemic dimension, she classified the dimensions of platforms from the four distinct perspectives. The first is through the technological lens, that is, the engineering design perspective. This eye sees the platform as technological architectures. The second is through the economic lens, that is, the economics perspective. This eye views the platform as markets. The third is through the organizational lens. This eye sees the platform as institutional-organizational elements. This lens enables us to see the firm structure, market structure, and regulatory structure of platforms.

Her final innovative view is through the epistemic lens. That is, the platform will bring new forms of knowledge into existence. It should be noted that this knowledge expanding function of platform is most important but rarely-treated in the literature. For example, the digital platform accompanied by transport services can provide the appropriate transportation service to the woman in the street, whatever her time, place, and state are. In textbook economics, this kind of knowledge is simply assumed to be given in our market economy. However, the economists such as Hayek, Arrow and Debreu insisted that the information on time, place, and state about the services or goods is very important in determining the values of commodities. While Arrow and Debreu identified the commodities (i.e., products and services) based on their characteristics such as its time, place, and states, Hayek (1945) emphasized the importance of the particular information on time and place in the functioning of market economic systems, compared with planning economy. Especially, Hayek insisted that the market economy with price system is an excellent and spontaneous mechanism to solve the social problem of how to secure the best use of resources, given that the knowledge of the circumstances is dispersed among the individuals and their knowledge is frequently contradictory.[3] Therefore, the introduction and widespread use of digital platforms make some enormous changes of these market mechanisms, which includes the determination of (relative) values of commodities and the market process itself.

Like Amazon and Uber, the platform providers are showing these knowledge expanding functions of platforms. Combined with network effects, this final element of platform also makes the platform revolution realized. In summary, technologies, markets, and institutions co-evolve (Table 2).

---

[3]Also, Hayek (2002) emphasized the discovery function of competition. These knowledge aspects of market and price systems have been more developed by so-called Austrian economics such as Kiesling (2018).

Table 2 Kiesling (2018)'s technology-economic-organizational-epistemic synthesis for platform

Technological lens, engineering design perspective: platform as technological architectures
(Explanation) A technological platform is a common core of technologies within a modular architecture, with variable technology elements around the periphery that interoperate with the core technologies and architectures
Economic lens, economics perspective: platform as markets
(Explanation) An economic platform is a means for facilitating and coordinating mutually-beneficial exchange through transactions in a two-sided market
Organizational lens: platform as institutional-organizational elements
(Explanation) An organizational platform can provide institutions that enable the coordination of the actions and plans of agents (be they individuals or firms) within a technology platform for mutual economic benefit, and it can have different organizational form in different industries and contexts. The organizational lens encompasses the three dimensions of an institutional framework: firms structure, market structure, and regulatory structure
Epistemic lens from Austrian Economics
(Explanation) Platforms are epistemic framework for decentralized coordination. Digital technology, multi-sided markets, and governance combine to take the technology-induced transaction cost reductions and create value by accessing diffuse private knowledge that was inaccessible in a pre-platform and creating new forms of exchange that lead to new knowledge generation in the discovery procedure. Platforms change the context of market activity in ways that bring new forms of knowledge into existence. For example, the knowledge of time, place, and states of the man on the street is relevant. In platforms, technologies make such diffuse private and contextual knowledge accessible and actionable. Prices and markets make more coordination possible than would happen in their absence, and platforms make more and different types of markets possible, deepening and broadening the scope of human activity over which we can achieve decentralized coordination and inducing experimentation and innovation

Source: Kiesling (2018)

3 Understanding Platform Structures in Container Shipping Markets

3.1 Step-Wise Process of a Typical Blockchain

For understanding the core functions of blockchain platforms in container shipping markets,[4] it is necessary to figure out the basic steps of a typical blockchain process, which is illustrated in Fig. 2. First, one man or woman in the P2P (person to person) network requests a transaction. Second, the requested transaction is broadcast to the P2P network consisting of computers (communication step), which is called as nodes. Third, the network validates the transaction and the users' status using some algorithm (validation step). Fourth, once verified, the transaction is combined with other transactions to create a new block of data for the ledger (verification step). Fifth, the new block is then added to the existing blockchain in a way that is permanent and unalterable (confirmation step). Therefore, the blockchain

[4]For more information on the blockchain in supply chain, refer to Wang et al. (2019).

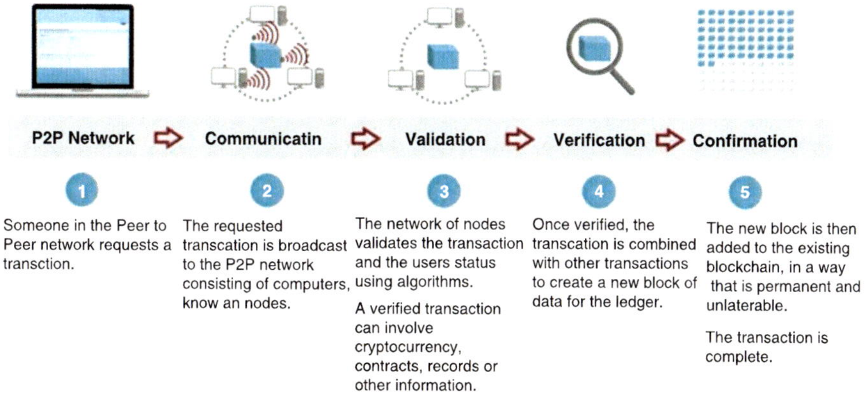

Fig. 2 Blockchain process steps. Source: Kommana (2017)

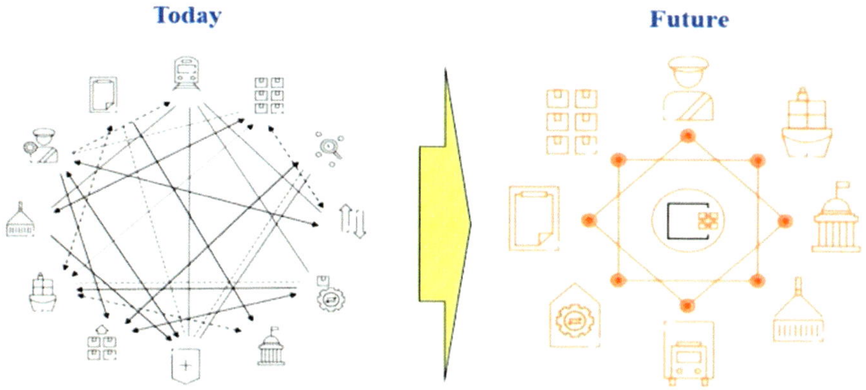

Fig. 3 Trade Network Transformation through TradeLens. Source: TradeLens (2019)

technology is evaluated as providing a decentralized platform, which can eliminate a number of risks associated with incumbent centralized system.

3.2 *TradeLens*

Figure 3 shows the pictures before-and-after the introduction of TradeLens. The participants in the international container transportation are typically export and import shippers, trucks, warehouses, port terminals, shipping companies, customs, port authorities, freight forwarders (or NVOCCs), and so forth. The transactions and documentations among and between these agents are very complex. TradeLens is a platform for these complex interactions. As of 2020, it is known that over 60% of market agents in global container shipping sectors are participating in TradeLens. If

TradeLens processes all these transactions and documentations, its effects and benefits will be enormous. First of all, TradeLens will enhance the transparency and security. Also, it will enable the more accurate forecasting for the future activities in container shipping sectors.

The mentioned benefits can come from the designed character of TradeLens. Figure 4 shows what are blockchained through TradeLens. All the transactions and documentations are verified by the simultaneous approvals of involved agents. These electronic records are protected from deletion, tampering, and revision by using the blockchain technology.

There is a powerful alternative blockchain platform against TradeLens in the global container shipping markets, that is, GSBN. GSBN is led by the Chinese container shipping lines such as COSCO Shipping. However, it should be noted that there are some market players, which simultaneously participate into GSBN as well as TradeLens, such as CMA-CGM, Hapag-Lloyd, and PSA. This multiplicity of platforms and the subsequent issue of optimal number of platforms is very important topic, which will be suggested as a future research agenda (Fig. 5).

While the TradeLens and GSBN are two distinct commercial blockchain-based digital platforms, there is a comprehensive non-commercial cooperating organization, Digital Container Shipping Association (hereafter DCSA)[5] that produces technical standards which are beneficial to all parties involved in international trade and aims to achieve the widest possible adoption of those standards. These standards establish the basis for system operability, which ensures the involved organizations are "speaking the same language". In other words, DCSA pursues to minimize barriers which may exist towards greater transparency and interoperability within the container shipping supply chain. Nine of the top 10 container companies are DCSA members. So, DCSA are representing approximately 70% of global container trade. Besides these container shipping firms, DCSA are collaborating with many parties such as from shippers, ports, terminals and freight forwarders to governments, regulatory bodies, banks and insurers.

3.3 FBX

Figure 6 shows the detailed information on the FBX (Freightos Baltic Global Container Index), which has a platform character. The FBX production firm, Freightos, is providing the market information services to 1600 logistics service providers and shipping carriers. Also, Freightos' platform consists of the three branch companies such as Freightos Markeplace, Freightos AcceleRate, and Freightos WebCargo. This large coverage, which might be enabled by digital technologies, gives some significant competitive advantage to the Freightos platform.

[5]The information on DCSA is from its website.

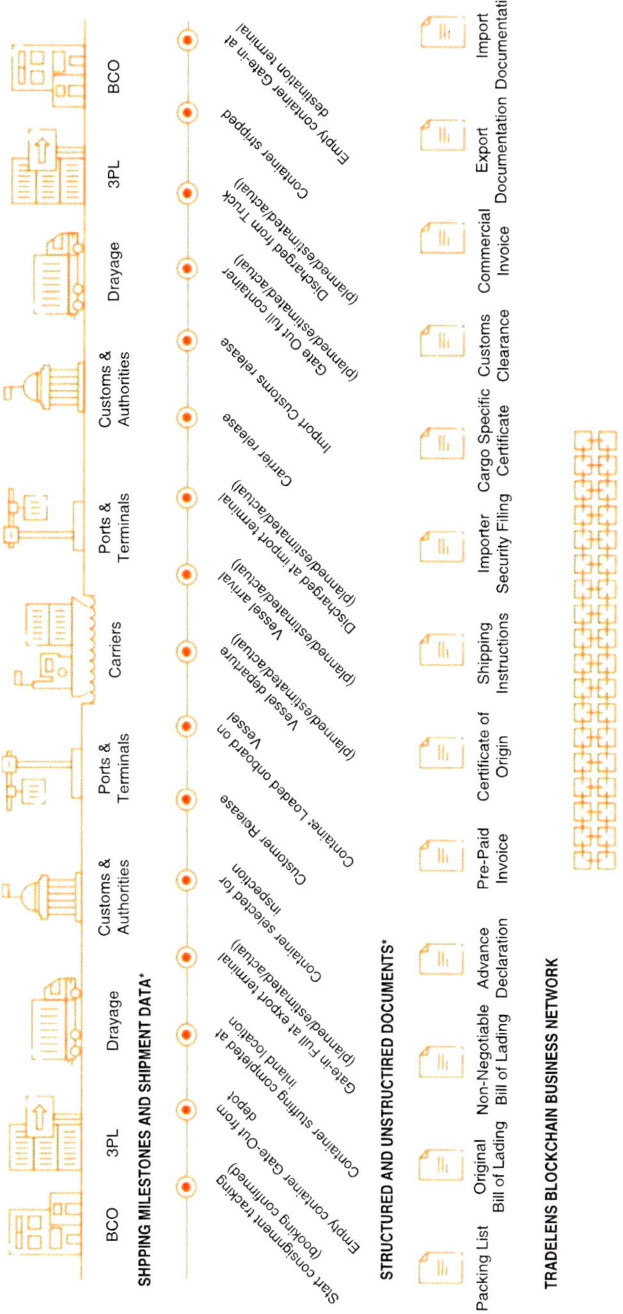

Fig. 4 What are block-chained by TradeLens. Source: TradeLens (2019)

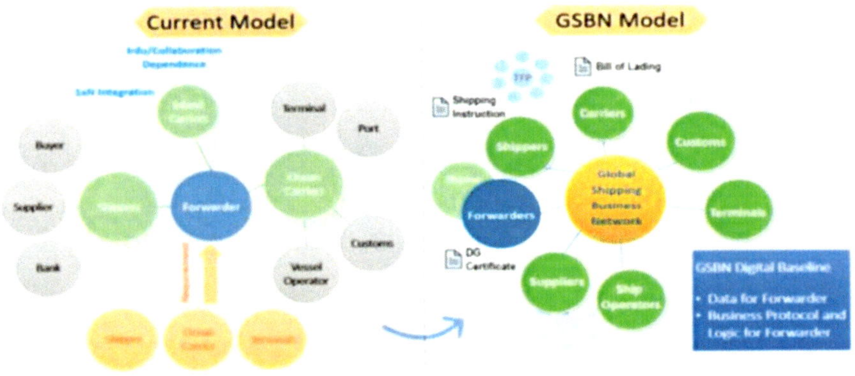

Fig. 5 GSBN Model. Source: CargoSmart AI

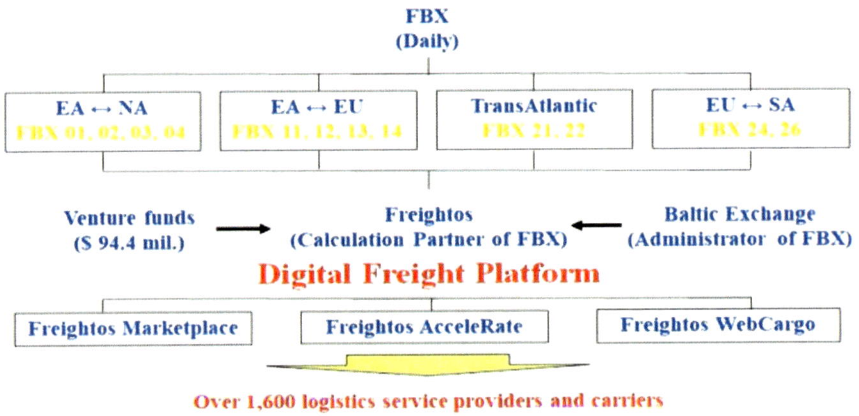

Fig. 6 Platform Structure of FBX. Source: Baltic Exchange (2019) and other press articles; Author

In addition, about 94 million dollars were invested to Freightos by venture capitalists. The indicator from Freightos is administrated by the Baltic Exchange. It is well known that the Baltic Exchange has long professional history in managing market indicators in bulk shipping markets such as BDI and its corresponding freight futures prices, i.e., Baltic Forward Assessment (BFA) for Forward Freight Agreement (FFA)'s clearing. In particular, FBX consists of the twelve global trunk routes, whose first four routes are transpacific, second four are East Asia-Europe, third two are transatlantic, and final two are north-south routes. The single composite index, FBX is published on daily basis. As of now, while only the single index, FBX is being published on daily basis, its sub-indices may also be published on daily basis in future, as the BDI and its sub-indices are published on daily basis. In addition, if the FBX-indexed transportation contracts and futures are used in larger scale, then the related futures prices can be introduced in the shipping markets as the BFA has been used.

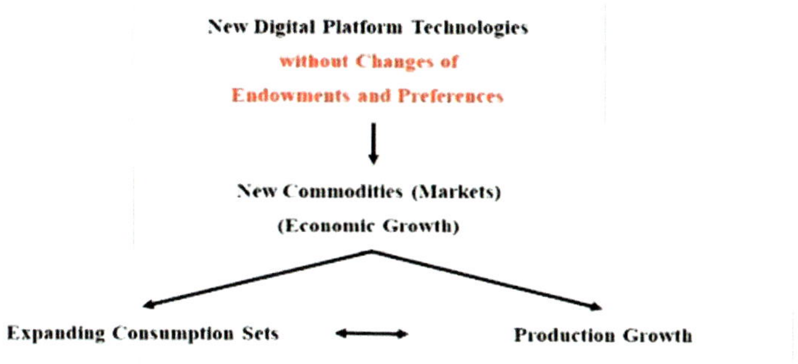

Fig. 7 General benefits from platform revolution. Source: Author

4 Prospective Benefits, Barriers and Threats

4.1 Benefits

Before considering the benefits of platforms in the shipping markets, we should consider general benefits from platform revolution. For discussing the benefits of the changes from platform revolution (e.g., emergence of the global e-commerce platforms such as Amazon), we need a theoretical framework. We could start from the so-called Walrasian general equilibrium model as a theory for the general market economy. The concrete model considered in this chapter is the Arrow-Debreu model.[6] Borrowing the conceptualization method of Balzer (1998),[7] we could state that, in the Arrow-Debreu's market systems, there are consumers, commodities such as products and services, firms, endowments such as natural resources and human's time, preferences over commodities, and production technologies. As a result, there would be general equilibrium states satisfying all the market participants.

Figure 7 shows the general benefits from platform revolution in a simple flow chart. The digital platform revolutions are now opening new possibilities without any change of the fundamental endowments and preferences. In particular, the newly-made exchange possibilities due to the lower transaction costs are now making new markets, that is, new commodities.[8] For example, e-commerce platform connects the consumers and producers who have not been matched by the traditional markets. In addition, the reductions of search and transportation costs make the decrease of related marginal costs and then make new transactions, which have not been possible in high marginal costs. This means that more products and services are

---

[6]In this chapter, we discuss Debreu (1959)'s theory as the Arrow-Debreu model.

[7]Balzer (1998) explains the set theoretic structuralism in economics. For this explanation, the Arrow-Debreu model was used as an illustrative example.

[8]Regarding the reductions of the economic costs such as search costs, replication costs, transportation costs, tracking costs, and verification costs from digital-intensive innovations, Goldfarb and Tucker (2019) documents well the previous economic literatures.

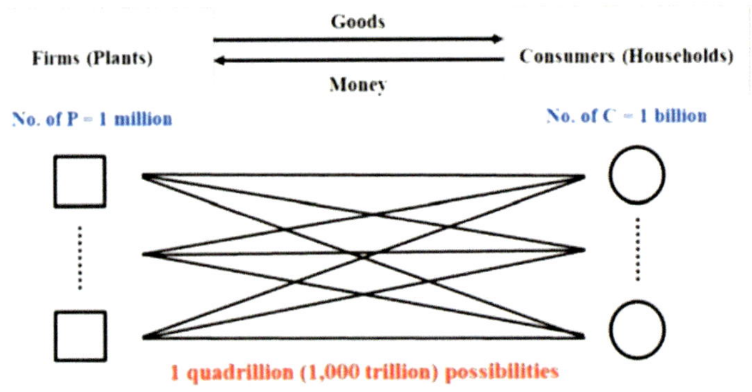

Fig. 8 Benefits from trade and logistics platforms. Source: Author

available to production firms and consumers. This also means that the related economies can grow more than before.

Using the theoretical framework of Arrow-Debreu's model, we can think the quantitative implications for the logistics industry from the platform revolution. Figure 8 shows my simple formulation. Suppose that there are 1 million firms (or plants) producing one kind of physical goods in our global economy. In addition, suppose that there are 1 billion consumers who have purchasing power. Simple calculation of the number of possible transactions yields the 1 quadrillion, that is, 1000 trillion possible transactions. The platform revolution such as the explosion of global e-commerce means that these huge possibilities can be realized. Simply imagine the functions of Amazon or Alibaba.

In short, these commercial trade platform companies are connecting most of consumers and producers in our global economy. A simple and powerful implication for logistics sectors is that we, logistics industry including shipping, should satisfy these potential transportation needs. It is important to note that, without international shipping services, these transaction possibilities cannot be realized in actual life.

Figure 9 shows the benefits from TradeLens (or GSBN). According to Cichosz et al. (2020), the current problems of logistics industry are high fragmentation, low transparency, underutilized assets, costly manual processes, and outdated customer interfaces. The platform solutions based on blockchain technologies will enable us to organize the logistics sectors in more innovative ways. Supported by blockchain technologies, global logistics industry will enhance the integration, transparency, utilization of assets, online processes and customer interfaces, simultaneously reducing the transaction costs in logistics sectors across global supply chains. This means that, in the logistics markets, there would be remarkable gains in the efficiency and effectiveness, which will result in the significant reductions of logistics costs. Therefore, due to these cost reductions, the global economy will grow more than before, that is, more expansion of markets.

Figure 10 shows the benefits from FBX, i.e., a platform-based container market indicator. FBX and its sub-indicators are new market representative indices. Using

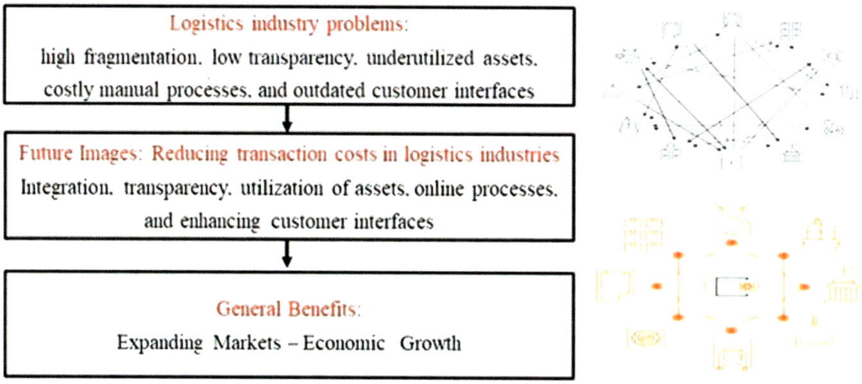

Fig. 9 Benefits from TradeLens. Source: (left) Author, (right) TradeLens

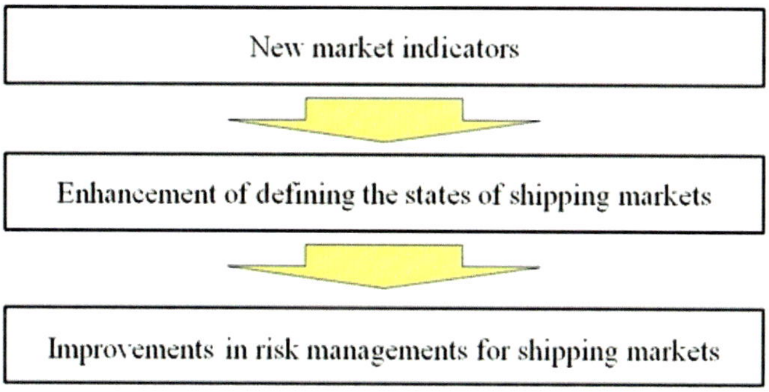

Fig. 10 Benefits from FBX. Source: Author

the terminologies of economic theory, we can enhance defining the states of shipping markets. This improvement in the definitions of market states means that we can design more contingent commodities for the market exchanges.[9] In other words, more accurate descriptions of the economic states enable the economic agents to define the future states more precisely in uncertain circumstances in advance and then this enhancement of defining future states will provide more risk management tools in various forms including risk sharing contracts.

For example, the famous futures markets such as FFA markets in shipping sectors could have been made possible by the use of Baltic Dry and Tanker indices.

[9]The concept of contingent commodities (or contingent contracts) was first introduced by Arrow (1953) and more developed by Debreu (1959). Mas-Collell et al. (1995), p. 687) defines a 'contingent commodity' as a commodity whose delivery is conditioned on the realized state of the world. Magill and Quinzii (1996) defines a 'contingent contract' for state s (s is one state of all possible states) is a promise to deliver one unit of the good in state s and nothing otherwise. The basic premise of this contingent commodities (or contracts) is that Nature determines the specific state from all the possible states and we, human, can define (or describe) their characteristics a prior.

Following this tradition, FBX will enable us to improve the risk managements in container shipping markets. For example, Drewry estimated that in 2018, over 10% of global container transportation contracts were so-called index-linked ones (Gore, 2018). Further, the chartering and ship financing contracts can be made based on these market indices. In particular, like the GDP-linked bonds or inflation-indexed securities, the shipping financial contracts can be indexed to these market indices, which, for example, will enhance risk sharing.

4.2 *Barriers*

Among various barriers to platform revolution, we should emphasize the importance of human resources. As well known, the cyber-attack and its prevention by security measures is also very important. However, all these activities must be conducted especially by the professional personnel. In particular, Cichosz et al. (2020) emphasized this fact by saying "people are the biggest challenge and main source of success to digital transformation" including platform revolution. Especially, for the success in responding to digital revolution, "there should be digital leaders with promising vision and these leaders should be supported by empowered, knowledgeable and collaborative employees".

4.3 *Threats*

The platform revolution brings out not only the benefits, but also the threats. As implied from a universal applicability of AI (e.g., see Sowa, 1999), in general there have been much concerns about the technological unemployment,[10] which occurs owing to the rapid substitution of human labor with automatic machines.

However, in this sub-section, I would like to explain the reason why the container shipping companies should transform into a digital-intensive organization by showing a potential scenario from the widespread use of AI (Artificial Intelligence) recommendation systems in the consumers' choice (see Fig. 11). AI uses some algorithm, which produces the choice recommendation based on the historical big data. For example, if the record of a shipping company cannot make the AI choose it as an alternative service provider, this tendency of rejection will persist, if without meaningful changes in the competitiveness of the relevant shipping firms. In other words, as well known, AI algorithm searches optimal patterns in the historical data for the decision makers. These optimal patterns are by nature a statistical regularity.

[10] Keynes (1930) defined technological unempoloyment as "unemployment due to our discovery of means of economising the use of labour outrunning the pace at which we can find new uses for labour". For more information on technological unemployment, refer to Wikipedia's service.

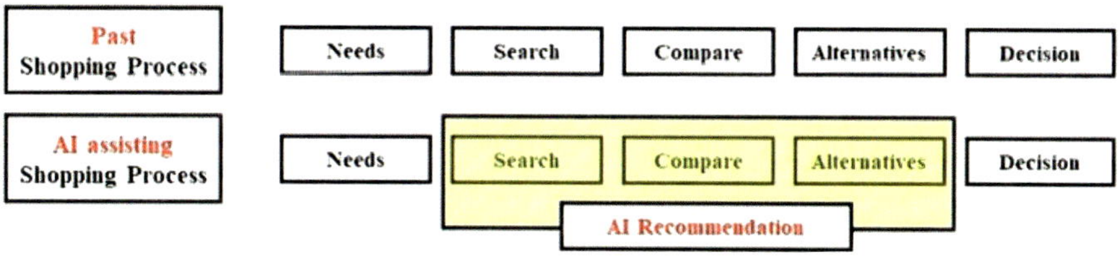

Fig. 11 Changes in shopping process by AI recommendation. Source: EY Korea (2020); Author's manipulation

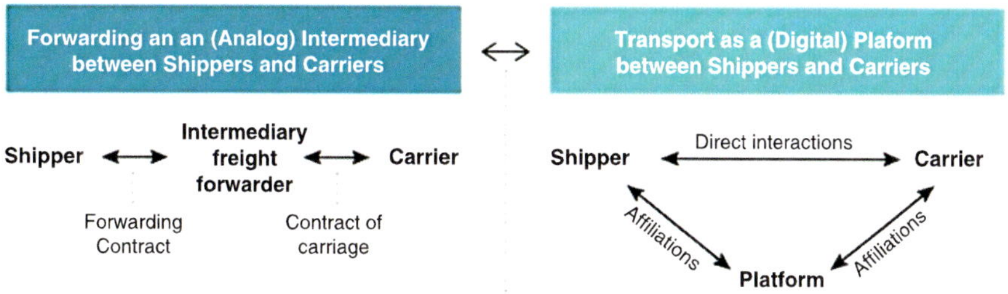

Fig. 12 Disintermediation by intermediating platform. Source: Deloitte (2019), p. 7

Therefore, if the accumulated data records regarding the related shipping firms exceed the AI's threshold for its exclusion choices, then it will become more difficult to enter again into the inclusion choices. That is, it is thought that there will be a serious lock-in effect in the alternative choices. In short, as the shippers use more AI-type automatic choice tool, the shipping companies should endeavor more to be selected by this AI algorithm.

In addition, as emphasized by Deloitte (2019) and EY (2017), there have been emerging serious concerns of disintermediation of freight forwarders, because of widespread use of platform-based transportation contracts. Figure 12 shows well these threats from the change into the disintermediation. This threat was that the freight forwarder will be redundant in supply chain activities. However, Deloitte (2019) revealed the survey results of industry-specific prohibitors that run counter to these concern. The prohibitors are (1) shipper's need for customization, guaranteed prices and loading capacity, and consolidated invoicing, (2) robust and rigid reliability requirements between different parties involved in transportation, (3) need for involvement of many different types of parties in shipments.

5 Conclusions

The main arguments of this chapter are as follows.

A platform is a form of intermediary using new digital technologies with the features of network and market. Without changes of the fundamental endowments

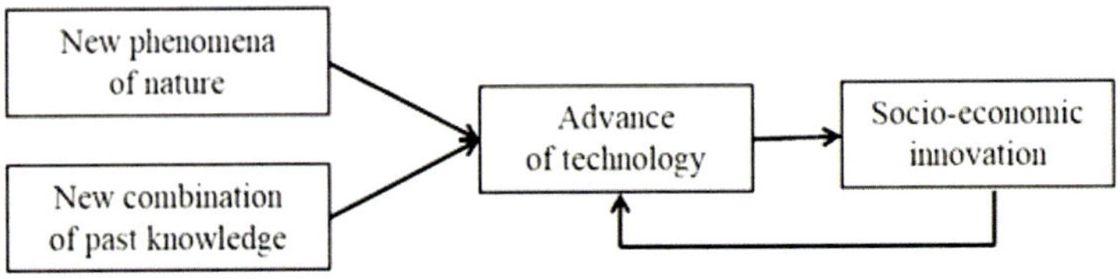

Fig. 13 Technology and innovation: a scheme. Source: Arthur (2009); Author's manipulation

Fig. 14 Consequences of platform revolution. Source: Author

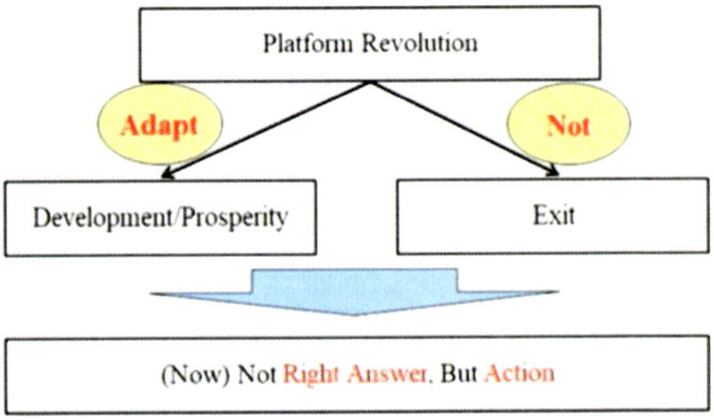

and preferences, new digital platform technologies are creating new markets and thus enabling economic growth. These benefits mean the expansion of consumption sets. Regarding the particular container shipping sectors, first, blockchain platforms are enhancing and will enhance the integration, transparency, utilization of assets, online processes, and customer interfaces in logistics, which results in remarkable gains in the efficiency and effectiveness of logistics sectors (i.e., reductions of logistics costs) and thus results in the expansion of markets (i.e., economic growth) in general. Second, new market indicators (e.g., FBX) will improve the market risk management in container shipping markets by the enhancement of defining the states of shipping markets.

According to Arthur (2009), as the discovery of new phenomena of nature and new combination of past knowledge enable the advance of technology, the associated socio-economic innovations can come out and the relevant economic systems can also develop. In this context, the 4th Industrial Revolution symbolizes the extremely speedy advances of technologies and their corresponding innovations (Fig. 13).

The subject of this chapter, platform revolution in container shipping markets, may be considered in these processes of economic evolution. The ultimate choices of consumers will determine the winners and losers. If the organizations such as firms adapt to these disruptive changes, they will develop and prosper. But, if they cannot adapt, they are forced to exit from markets. In other words, it is the swift practical 'action', but not the 'right answer', that is necessary for the competing firms (Fig. 14).

There are several important future research topics, which are not treated in this chapter. Among them, I would like to mention the four agendas.

First, I think that there is a big research opportunity on the topic of digitalization and digital transformation[11] in container shipping markets, which is deeply interrelated with digital platform revolution, the topic of this chapter. Regarding the digital transformation of the economy in general, Hanna (2016) and Westerman et al. (2014) provide excellent information. In particular, OECD (2019) and EU (2018) suggest the methods of measuring the digital transformation in the relevant economies. However, to the best of my knowledge, there have rarely been comprehensive documents or researches on the digital transformation in the container shipping markets and its measuring methods, which will be worthy of future research efforts.[12] In addition, this disruptive development of digitalization will require a newer policy approach. Among the efforts of policy makers for this direction, it is notable that Patton (2011) argues that 'developmental evaluation' can be an excellent candidate for policy evaluation and thus policy design (making).

The second is on the diversity in the blockchain platforms in container shipping markets. As explained, in container shipping markets, there are two alternative blockchain platforms, that is, TradeLens and GSBN. If we think only the efficiency in these platforms, due to the size and network effects, only one platform is desirable. However, this degree of optimal diversity is not simple as seen as superficially. So, I think that it is necessary to study this topic in the perspective of industrial organization of economics.

The third topic is about the use and activation measures of index-linked contracts in container shipping markets. In fact, before the introduction of FBX, there have been well-designed market indicators in container shipping markets such as SCFI and Drewry Index. However, there have been some yet-not-studied impediments to the success of these indicators as risk management tools.[13] So, I think that for the success of platform indicators such as FBX, our maritime economists should conduct the research on the success measures for it.

The last, but not the least topic is the micro analysis of the benefits from and barriers to the platform revolution in container shipping markets. Up to now, this article has focused on the industry- or macro-level origins and characteristics of

[11]Morakanyane et al. (2017) defines the digital transformation is "an evolutionary process that leverages digital capabilities and technologies to enable business models, operational processes and customer experiences to create value".

[12]In 2019, Google's quantum processor, named Sycamore, performed a computation in 200 seconds that would have taken the world's fastest supercomputer 10,000 years to solve. With regard to this quantum computing, FreightWaves (2020.6.22) reported that SavantX with D-Wave's quantum services has begun offering its technology to solve optimization problems to customers such as the Pier 300 container terminal project at the Port of Los Angeles. This kind of movement implies that there will be much more impact from the development of digital technology (digitalization) in general and in container shipping sectors.

[13]According to the estimate of Drewry, in 2018, about 10% of contracts in container shipping transportation were index-linked.

platform revolution. However, the firm- and consumer-level study (i.e., micro analysis) is required, especially for the decision makers of private sectors.

Acknowledgements The author would like to express deep gratitude to President Dr. Young-Tae Chang of Korea Maritime Institute and Prof. Dong-Wook Song of World Maritime University, a co-editor of this book, for their invaluable encouragements and advices for this article.

References

Arrow, K. J. (1953). Le role des valeurs boursieres pour la repartition la meilleure des risques. In *Econometrie, Colloques Internationaux du Centre National de la Recherche Seientifique* (pp. 41–47).; English version: "The role of securities in the optimal allocation of risk bearing," *Review of Economic Studies*, 31, pp.91-96, (1964).

Arthur, W. B. (2009). *The nature of technology: What it is and how it evolves*. Free Press.

Baltic Exchange. (2019). *Guide to Freightos Baltic Global Container Index (FBX)*, Version 1 – December 2019, Baltic Exchange.

Balzer, W. (1998). Set theoretic structuralism. In J. B. Davis, D. W. Hands, & U. Mäki (Eds.), *The handbook of economic methodology*. Edward Elgar Publishing Limited.

Cichosz, M., Wallenburg, C. M., & Knemeyer, A. M. (2020). Digital transformation at logistics service providers: Barriers, success factors and leading practices. *The International Journal of Logistics Management, 31*(2), 209–238.

Debreu, G. (1959). *Theory of value: An axiomatic analysis of economic equilibrium*. John Wiley & Sons, Inc./Chapman & Hall, Ltd.

Deloitte. (2019). *Digitalization in freight forwarding – Beyond the platform hype*. Deloitte.

EU. (2018). *Digital transformation scoreboard 2018: EU businesses go digital: opportunities, outcomes and uptake*. European Union.

EY. (2017). *Disruption. digitalization. dsintermediation: Transportation and logistics in the coming decade*. EYGM Limited.

EY Korea. (2020). Super consumer. In *Korean language*. EY Korea.

FreightWaves. (2020). *Tech company uses quantum computers to help shipping and trucking industries*. FreightWaves.

Gawer, A. (2014). Bridging differing perspectives on technological platforms: Toward an integrative framework. *Research Policy, 43*, 1239–1249.

Goldfarb, A., & Tucker, C. (2019). Digital economics. *Journal of Economic Literature, 57*(1), 3–43.

Gore, M. (2018). *Index linked container contracts – The future in a volatile market?*.

Hanna, N. K. (2016). *Mastering digital transformation: Towards a smarter society, economy, city and Nation*. Emerald Group Publishing Limited.

Hayek, F. A. (1945). The use of knowledge in society. *American Economic Review, 35*(4), 519–530.

Hayek, F. A. (2002). "Competition as a discovery procedure" translated by Snow, MS. *The Quarterly Journal of Austrian Economics, 5*(3), 9–23; The Original lecture in German language was delivered in 1968.

Keynes, J. M. (1930). *The economic possibilities of our grandchildren*.

Kiesling, L. (2018). Toward a market epistemology of the platform economy. In *Austrian economics: The next generation* (Vol. 23, pp. 45–70). Emerald Publishing Ltd.

Kommana, B. (2017). *Blockchain: Moving beyond bitcoin*. MSG-global.

Magill, M., & Quinzii, M. (1996). *Theory of incomplete markets*. Massachusetts Institute of Technology.

Mas-Collell, A., Whinston, M. D., & Green, J. R. (1995). *Microeconomic theory*. Oxford University Press, Inc.

Morakanyane, R., Grace, A. A., & O'Reilly, P. (2017). Conceptualizing digital transformation in business organization: A systemic review of literature. *Bled eConference*, 427–443.

OECD. (2019). *Measuring the digital transformation: A roadmap for the future*. OECP.

Parker, G. G., Van Alstyne, M. W., & Choudary, S. P. (2016). *Platform revolution: How networked markets are transforming the economy and how to make them work for you*. W. W. Norton & Company.

Patton, M. Q. (2011). *Developmental evaluation: Applying complexity concepts to enhance innovation and use*. The Guilford Press.

Rysman, M. (2009). The economics of two-sided markets. *Journal of Economic Perspectives, 23* (3), 125–143.

Sanchez-Cartas, J. M., & Leon G. (2019). "*Multisided platforms and markets: a literature review.*", unpublished manuscript accessed via ResearchGate.

Sowa, J. F. (1999). *Knowledge representation*. Course Technology Ptr.

TradeLens. (2019). *TRADELENS*, Pusan.

Wang, Y., Han, J. H., & Beynon-Davies, P. (2019). Understanding blockchain technology for future supply chains: A systematic literature review and research agenda. *Supply Chain Management: An International Journal, 24*(1), 62–84.

Westerman, G., Bonnet, D., & McAfee, A. (2014). *Leading digital: Turning technology into business transformation*. Harvard Business Review Press.

Liner Services in Asia's Major Container Ports: Implications for Busan

Satya Sahoo

Abstract The international trade of commodities is crucial for the economic development of any nation or region. Over 80% of the international trade of commodities takes place through ships. Hence seaports serve as the gateway to the development of any country. Since Asia is the world's manufacturing hub, this chapter investigates the trading pattern of container line services for three major ports in Asia: Busan, Singapore, and Shanghai. As container lines provide services in closed loops, multiple ports are called during the services. Additionally, ports compete for common market share as there are overlaps in ports' hinterland and foreland market. Understanding the existence of co-operation and competition between ports (Song *Journal of Transport Geography* 10: 99–110, 2002), this study proposes a simple port intra-dependency ratio to measure the degree (or percentage) of which a particular port depends on reference port(s) in the neighbourhood. The study utilizes the automatic identification system (AIS) and liner services data from AXS Marine's Alphaliner for the analysis. Results suggest that the Port of Shanghai generates almost 26% of its volume without depending on Busan or Singapore port. Singapore port gets almost 40% of the liner capacity without depending on Busan and Shanghai, while Busan port generates only 11% liner capacity on its own, that is, without relying on Singapore and Shanghai. This indicates that the Port of Busan is heavily dependent on Singapore and Shanghai port for its container services, increasing its risk profile. Busan needs to diversify the liner services/routes portfolio and expand its self-dependency to become the transhipment hub in the Far East region. This study concludes by identifying some challenges and opportunities for the Port of Busan as a case study. Big data in maritime can act as a decision support system for port managers and policymakers to uncover the market's gaps and needs, thus increasing the port market share and retaining the high container throughput.

S. Sahoo (✉)
World Maritime University, Malmö, Sweden
e-mail: srs@wmu.se

B.-W. Ko, D.-W. Song (eds.), *New Maritime Business*, WMU Studies in Maritime Affairs 10, https://doi.org/10.1007/978-3-030-78957-2_9

1 Introduction

International trade of commodities is crucial in transforming the economy of a country or a region (Shafaeddin, 1995). To demonstrate it with an example, countries producing a large volume of steel need a large volume of iron ore and coal. Steel is used for infrastructure and manufacturing (such as automobiles, real estate, transportation) industry. Hence, a country importing a large volume of iron ore and coal anticipates booming of the country's infrastructure and manufacturing sector, increasing its gross domestic production (GDP). The growth in the manufacturing and production sector leads to a rise in the export of finished goods. As there is value addition in converting the raw material to finished goods, the balance of trade for a country increases with an increase in import of raw materials and export of finished goods (Hume, 1752), leading to a rise in the GDP of the country. Hence, monitoring the import and export of raw materials and finished goods worldwide can help predict the country/region's economic growth.

The shipment of commodities around the world takes place via three primary modes of transportation: (1) air—using commercial cargo aeroplanes; (2) land—using trucks for road transportation or cargo trains for rail transportation; and (3) water—using ships for seaborne trade. The seaborne trade through international ocean-going ships is responsible for around 80% of the international trade of commodities (Hoffmann et al., 2018) as it offers reliable and cost-effective transportation of goods. In fact, the cost of seaborne trade (referred as the freight rates) relative to the value (or price) of goods and household income level has reduced over time (Ma, 2020), indicating that the transportation of goods via ships has relatively become cheaper. As presented in Fig. 1, the transportation costs of a forty-foot equivalent (FEU)[1] container box between Asia—North America (trans-pacific trade route) has reduced from 30% of US per capita GDP in 1968 to only 3% in 2018. Hence, shipment capacity for an individual household has increased compared to the past, surging the trading activities.

The international trade of commodities and goods (exports) has increased from 130 billion US$ in 1960 to over 16 trillion US$ per year in 2016. Along with the increase in the value of the trade, as shown in Fig. 2, the world merchandise trade also increases to over 20% of the world GDP in 2016 compared to only 10% of the global GDP in 1960. This also indicates that the contribution of the international trade of commodities towards countries' or regional economy have gradually increased; hence it needs more attention than ever before. The global nature of this industry makes it vulnerable to the risks and uncertainties around the world. For example, the financial crisis in 2007-08, BREXIT in 2016, geopolitical tension resulting in the closing of the Suez canal in 1967 to the ongoing COVID-19 pandemic has an impact on the maritime and international trade. To explain with

[1]FEU is one of the standard size of container box used in the Asia—North America trade.

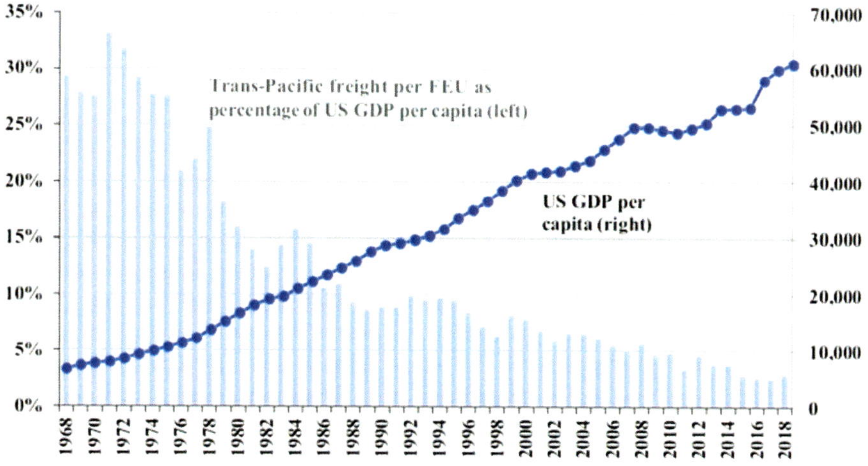

Fig. 1 Freight Rate vs US GDP per capita. Source: Ma (2020) compiled using Llyod's List, Clarksons and the US Census Bureau

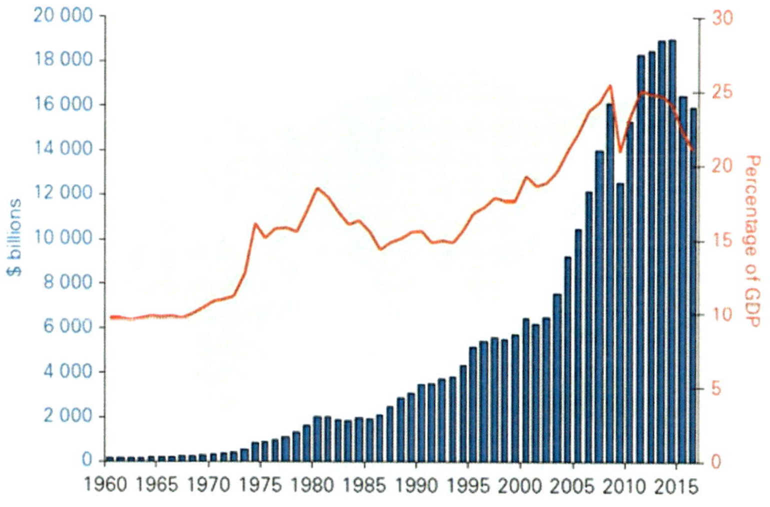

Fig. 2 World merchandise exports and percentage of world GDP. Source: Sloman et al. (2018) using data from the World Trade Organisation (WTO) and the World Bank

an example, during the current COVID-19 pandemic, the risk of breakdown of the global supply chain has increased, which could result in a shortage of essential commodities such as food and medicines. Therefore, the importance of international trade of goods and commodities is appreciated by practitioners, policymakers, and even the general public as a whole. Overall, the international trade of commodities not only comprises a significant fraction of the regional and global GDP but also affected by almost anything and everything happening around the world.

Since maritime trade comprises of 80% of the international trade of commodities, the seaports (hereafter referred as ports) holds two primary responsibilities: (1) acts as a node connecting the seaborne transportation to hinterland connectivity, and (2) serves as the gateway to the regional economy (Jung, 2011; Park & Seo, 2016). Ports not only facilitate exports and imports of raw material and finished (containerized) products which indirectly enhances the regional economy, it also provides logistics value-added services and employments in the region, which directly boosts the local economy. Hence ports (especially container ports) need to be strategically built and operated to support the country's economy. In the past, the port and maritime data were quite secretive and not readily available. So, even though ports are vital for any nation, ports' strategic decisions were not entirely scientific and somewhat depending on qualitative information. In the current era of digitalization and big data, information related to maritime and international trade of commodities is available at ease. That's why it becomes paramount importance for a port company to gather operation and business-related information for its own as well as for its competing ports.

Due to the development of digitalization and availability of big data in maritime and ports, this chapter aims to provide an overview of the container liner services and trade for three major ports in Asia: (1) Busan, (2) Singapore, and (3) Shanghai. Asia is the container trading hub of the world with almost half of the world container ships are deployed to provide liner services connecting the region to the world. Further, as 9 out of the top 10 container ports of the world are located in Asia, understanding the services for the major Asian ports can be interesting for both liner and port companies globally. *Secondly*, this study also attempts to understand the intra-dependency between the container ports to address the degree of co-operation and competition between the ports. *Finally*, this study identifies various challenges and opportunities using Busan port as a case study.

The remaining section of the study is presented as follows: Sect. 2 describes the relevant literature, identifies the gaps and highlights contribution. Preliminary discussion of data and the method used for the analysis is presented in Sect. 3. Section 4 presents the findings of the study, along with an in-depth discussion. The challenges and opportunities for Port of Busan as a case study are presented in Sect. 5. This section also provides some qualitative recommendations from the findings to improve the market share for Busan port. Finally, the study is concluded in Sect. 6.

2 Literature and Contribution

The list of the top 10 container ports in the world is presented in Table 1. It is interesting to note that 9 out of the top 10 container ports in the world are solely located in three Asia countries: China, Singapore and South Korea.

Since mega ports in Asia are mainly located in China, Singapore and South Korea, this study investigates the container trading pattern, co-operation and intra-dependency of the leading ports in each of the three countries, that is Shanghai port,

Table 1 List of top 10 container ports by throughput in 2019

Ranking	Port	Country	Region	2019 annual throughput (TEU)	2018 annual throughput (TEU)	Annual % change
1	Shanghai	China	Asia	43,303,000	42,010,200	3.1%
2	Singapore	Singapore	Asia	37,195,636	36,599,300	1.6%
3	Ningbo-Zhoushan	China	Asia	27,530,000	26,351,000	4.5%
4	Shenzhen	China	Asia	25,770,000	25,740,000	0.1%
5	Guangzhou	China	Asia	23,236,200	21,922,100	6.0%
6	Busan	South Korea	Asia	21,992,001	21,662,572	1.5%
7	Qingdao	China	Asia	21,010,000	19,315,400	8.8%
8	Hong Kong	China	Asia	18,361,000	19,596,000	−6.3%
9	Tianjin	China	Asia	17,264,000	15,972,000	8.1%
10	Rotterdam	The Netherlands	North Europe	14,810,804	14,512,661	2.1%

Source: Lloyds Maritime Intelligence

Singapore port and Busan port.[2] To understand the intra-dependency of the ports, it is essential as well to assess the trading pattern (export and import) of commodities for the corresponding countries. Table 2 represents the top 10 trading partner countries by the value of export and import for South Korea, Singapore and China. It is observed that South Korea majorly trades with China, USA and Japan. It also has other Asian and European countries as strong trade partners. Similar to South Korea, China and the USA are the leading trade partners for Singapore. Singapore also presents a good trade relationship with other Asian countries, mainly from the South Asian region. The value of goods/commodities exported and imported from China is significantly higher than that of South Korea and Singapore. Unlike the two countries who have China as the major trade partner, the USA is the major trading country for China. China also has a strong trade relationship with the Far East countries (including Japan and South Korea), South Asian countries such as Vietnam and Malaysia and with Europe. The differences in import and export of commodities create the balance of trade[3] for any country and establish dependency on each other. This dependency of countries on each other for trade creates resilience to find another trading partner, which is reflected on their corresponding trading ports creating co-operation and intra-dependency between ports.

[2]The ports are sometimes referred by their city names in the texts for easier reading experience; for example as Busan instead of Port of Busan or Busan port.

[3]The following link presents interesting racing bar chart of top importing and exporting countries of the world since 1992 along with top trade surplur and deficit countries: https://youtu.be/9YpwzM4DShQ.

Table 2 Top 10 trading partners for South Korea, Singapore and China

	South Korea		Singapore		China	
	Partner Countries	Value of trade	Partner Countries	Value of trade	Partner Countries	Value of trade
1	China	268.61	China	100.03	USA	635.71
2	USA	132.12	Malaysia	87.67	Japan	327.64
3	Japan	85.13	USA	73.97	Rep. of Korea	313.60
4	Viet Nam	68.27	China, Hong Kong SAR	52.30	China, Hong Kong SAR	311.52
5	China, Hong Kong SAR	47.97	Other Asia	48.38	Other Asia	226.00
6	Other Asia	37.52	Indonesia	48.16	Germany	184.17
7	Australia	30.33	Japan	42.14	Australia	152.63
8	Saudi Arabia	30.29	Rep. of Korea	29.85	Viet Nam	148.10
9	Germany	30.22	Thailand	23.85	Brazil	110.87
10	Russian Federation	24.82	India	19.59	Malaysia	109.17

Source: UN Comtrade
Note: Value of trade is comprising of total export and import in 2018, denoted in Billion USD

The co-operation and competition amongst regional ports are well investigated in the literature. Song (2002) examines the existing competition and co-operation between the container ports in Hong Kong and South China from a strategic point of view. The impact of such competitive strategies is now observed with the continuous decline of the container throughput for Hong Kong[4] and the growth of the adjacent ports in South China. In the early 1990s, Hong Kong was the largest container port in the world, but after 1997, when its sovereignty was handed to China, it faced steep competition from neighbouring Chinese ports, notably Shenzhen and Guangzhou. Using a game theory approach Wang et al. (2012) identifies several factors such as regional port governance for supporting co-operation and alliance formation to attract container volumes from contestable hinterlands. It also helps to develop an efficient logistics hub with a fully integrated transport ecosystem. Anderson et al. (2008) also extended the study of port competition from a game-theoretical perspective to understand how competitor ports respond when a neighbourhood port grows faster to become the regional container hub port. It also presents whether and how the growing port captures and defends market share by building additional infrastructure. The study also investigates the game theory framework using a case study between Busan and Shanghai port.

Though the concept of competition and co-operation between ports are extensively investigated, there are only limited studies which quantify the degree of competition and co-operation for ports. Asgari et al. (2013) is one of the few articles which empirically investigates the competition and co-operation between two major hub ports in Asia (Singapore and Hong Kong) using various scenarios. The current

[4]In 2019, the container throughput for Hong Kong port declined by 6.5%.

era of technology allows to digitize the operations, store huge information and process to useful findings almost instantaneously. The live and historical position of ships are tracked through AIS, stored and processed through various cloud platforms. The technology has also become cheaper and more predominant over time. Hence the usage of big data in the maritime industry is also gaining popularity. This study not only provides a simple empirical way to estimate the port dependency but also helps to understand the trading pattern of container services for major ports in Asia using liner service trade data that can enhance the management decision making for ports, container lines, and even regional governments (Sena et al., 2019; Tan et al., 2017).

3 Method and Data

As discussed in the literature review section, unlike tramp (dry bulk and tanker) shipping, container shipping is liner services with fixed (closed) routes and port calls. Since the liner services do not receive all their container boxes (either for loading or discharging) from a single port in the region, they rely on several ports in the region for their volume. For better understanding lets consider a scenario: say for liner services between Far-East and North-Europe route. Typically, container services load boxes in Asia and unload the boxes in Europe as Asia (mostly China) is the manufacturing centre of the world due to lower labour costs and well-established production ecosystem and Europe is the consuming centre due to higher per capita GDP. If we observe the loading centre for this service routes, not all the filled containers are loaded from China. There are other major manufacturing countries like Indonesia, Malaysia, Vietnam, Thailand, Bangladesh and India, amongst others, who either provide cargo volume to the Europe-Asia route directly or through transhipment ports such as Singapore and Busan. Furthermore, there are multiple mega ports[5] within China which also produces their cargo volumes and attracts the liner services. From a liner service provider perspective, since it is not possible to get the entire volume of cargo from a single port, it is ineffective to call only one port in the region, rather a loop of ports that meets its container volume requirement. Hence regional/adjacent container ports need to cooperate amongst each other for attracting container liner services. Additionally, ports strategically develop value-added services complementing the adjacent ports and establish a logistics ecosystem. Similar co-operation strategies are also designed for the regional ports at the discharge area, that is, for European ports in this example.

Regional ports not only cooperate but also compete to get a larger volume of cargo. Usually, this competition for cargo volume is for transhipment containers and not for hinterland containers since port generating their hinterland volumes have fewer options to share with other regional ports. Transhipment containers increase

[5]Currently, 7 out of top 10 container ports are in China.

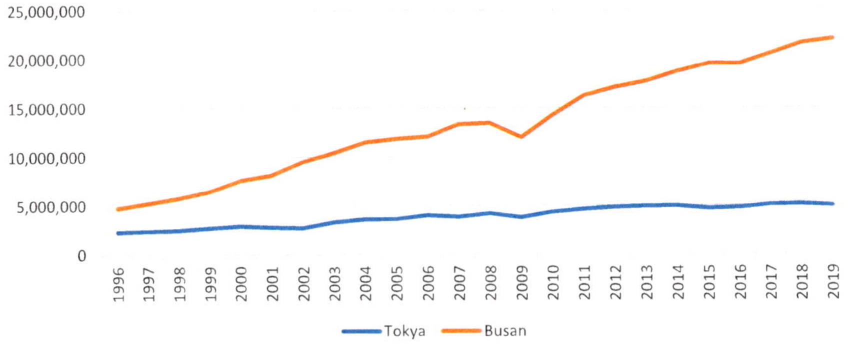

Fig. 3 Annual container throughput for Port of Tokyo vs Port of Busan. Source: Clarkson's Shipping Intelligence Network

the throughput volatility of a port (Notteboom et al., 2019), hence are quite risky for the ports. Ports with higher transhipment volumes compete with the adjacent ports to retain and increase their volume, which otherwise can significantly reduce their throughput volume, leaving them out of the mainline routes. As shown in Fig. 3, in 1996, Port of Tokyo and Port of Busan had annual throughput of 2.3 and 4.7 Million TEUs respectively; that is, Tokyo's throughput was almost half that of Busan. In 2019, the throughput of Tokyo and Busan was 4.9 and 22 million TEUs respectively; indicating that Port of Tokyo receives a container volume less than 1/4th of Port of Busan. Even though both the ports are located in the same region, by attracting large volumes of transhipment containers, the Port of Busan has emerged as the leading port in Far East Asia.

As regional ports cooperate for their existence and compete for scaling up their size, they need to develop a strategy to balance the co-operation and competition for a win-win scenario between ports (Hales et al., 2016). Brandenburger and Nalebuff (1995) introduced the word co-opetition, referring to a game theory strategy where players cooperate and compete. This is also in-line with the context of port co-operation and competition by Song (2002). Though port co-operation is quite essential, there are only limited studies empirically evaluating the degree of co-operation or dependency between ports. This study attempts to present the port intra-dependency ratio to quantify the dependence of a port with adjacent/reference ports as follows:

$$R_i^{\sum j} = \frac{TEU_{\sum j}}{TEU_i} \tag{1}$$

Where, $R_i^{\sum j}$ represents the port intra-dependency ratio of port i on the adjacent/reference port(s) j which are using the common liner services. TEU indicates the

Table 3 Comparison of container liner services: World vs Major Asian Ports

	World container trade	Busan	Singapore	Shanghai
No. of Services	1842	261	334	374
No. of Vessels Deployed	6136	944	1247	1761
Avg. TEUs capacity per week	3,600,134	925,332	1,340,815	1,525,531

average weekly 20-foot equivalent container capacity calling the port.[6] The ratio ($R_i^{\sum j}$) estimates the percentage of the volume of cargo (in the form of TEU) which a port i depends on the reference port(s) j compared to its total volume. A higher (lower) $R_i^{\sum j}$ denotes higher (lower) dependency of port i on reference port(s) j. This port intra-dependency ratio is an augmented form of port dependency ratio presented in DeSalvo (1994) and Won et al. (2006) which is derived from the demography and health economics literature (Loichinger et al., 2017; Muszyńska & Rau, 2012; Sinnathurai, 2013).

3.1 *Data and Description*

The data related to container liner services and the vessels allocated for the services are collected from AXSMarine's Alphaliner database. Alphaliner combines individual vessels allocated for each container liner services with the automated identification system (AIS) of the ship to provide updated routes and port calls.[7]

There are 6136 fully cellular container ships providing 1842 container liner services globally. Three thousand and sixteen container vessels are responsible for 767 liner services associated with Asia, either trading within the continent or connecting Asia to the world. Within Busan, Singapore and Shanghai, Shanghai is connected with the largest number of services, followed by Singapore and Busan respectively. Similarly, 1761 vessels are deployed to provide the liner services to Shanghai, followed by Singapore with 1247 vessels and Busan with 944 vessels. Over 1.5 million cumulative TEU capacity calls Shanghai every week, whereas Singapore and Busan receive 1.34 million and 925 thousand TEU capacity weekly. The details of the number of services, vessels allocated, and weekly TEU capacity for the ports are provided in Table 3.

[6]Please note: TEU does not refer to the actual weekly continer turnover volume (loaded and unloaded) of the port, rather the cumulative 20-foot equivalent container space in fully cellular ships calling the port on a weekly basis. The actual volume of container turnover is only available to the ports and terminals which is not shared with other. Further, as the study considers the ratio, rather than the absolute value, the cumulative volume of container space can reflect the dependence of ports.

[7]The container liner services for October 2020 is used; the results might be different at a different period of time.

4 Findings and Discussion

As discussed earlier, container liner services are operated in a close loop, where the liner services call multiple loading and unloading ports. A total of 261 container liner services with a cumulative capacity of 925,332 TEUs call Busan port every week. At the same time, Singapore and Shanghai receive a cumulative capacity of 1,340,815 and 1,525,531 TEU from 334 and 374 container liner services weekly. There are 54 common container services between Busan, Singapore and Shanghai port which provides a total of 457,859 TEU capacity weekly. There are eight liner services common between Busan and Shanghai exclusively[8] while Shanghai shares a high number of common liner services (83 and 64 respectively) with both Busan and Singapore. This indicates that Busan and Singapore do not depend on each other exclusively for getting the container volume while both the ports depend on Shanghai for getting their volumes. There are only 116 liner services exclusively calling Busan (that is, not calling Singapore and Shanghai) getting a cumulative container capacity of 104,957 TEUs weekly. This is lower than 1/9th of the weekly TEU capacity, indicating that, Busan heavily relies on Singapore and Shanghai for getting its cargo volume. On the other side, Singapore gets almost 531,189 TEU capacity weekly from 208 liner services which do not call Busan and Shanghai, which is nearly 1/3rd of its total TEU capacity, while Shanghai receives about 1/4th (402,355 out of 1,525,531) of its TEU capacity on its own. Shanghai has a large hinterland for which it can self generate higher cargo capacity without depending on the other two ports. In contrast, Singapore, despite being a transhipment port, generating a higher TEU capacity without relying on adjacent ports (Busan and Shanghai in this case) indicates that it has a well-diversified network of liner services. The container market share for Busan, Singapore and Shanghai is graphically presented using a Venn diagram in Fig. 4.

As seen in the Venn diagram, the entire container market share for Busan, Singapore and Shanghai can be divided into three zones as follows:

1. Common liner services for three ports: This includes liner services that call Busan, Singapore, and Shanghai.
2. Liner services for two ports only: This includes the liner services calling only two ports, that is, (1) Busan—Singapore, (2) Busan—Shanghai, or (3) Singapore—Shanghai.
3. Liner services only calling one port: Out of the three ports considered in the analysis, this includes the liner services calling only Busan, Singapore, or Shanghai.

[8]Exclusively here refers to the liner services which are only calling the aforemented port(s) and not with the port(s).

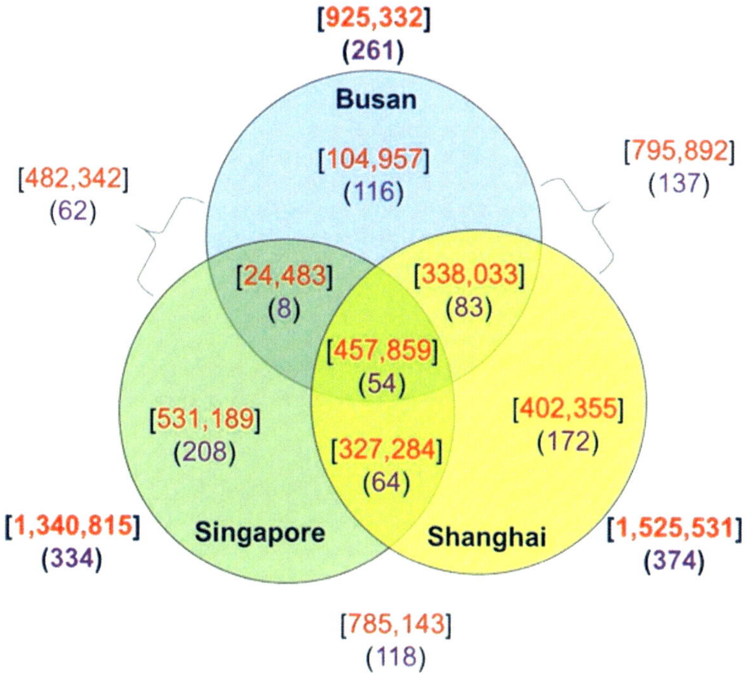

Fig. 4 Container liner market share. Note: [XX] represents avg. TEUs capacity per week while (XX) represents the number of services

Based on the three aforementioned zones for liner services, the intra-port dependency ratio is evaluated and explained in the following section.

4.1 *Common Liner Services for Busan, Singapore, and Shanghai*

The port intra-dependency ratio ($R_i^{\sum j}$) for Busan port with respect to adjacent all three ports (that is, port *i* is Busan while ports *j* are Busan, Singapore and Shanghai) is 49.48%. This indicates that Busan depends on both Singapore and Shanghai port for almost 50% of its volume. Similarly, the port intra-dependency ratio ($R_i^{\sum j}$) for Singapore and Shanghai port is 34.15% and 30.01% respectively suggesting that Shanghai depends list on Busan and Singapore for generating its volume.

Container liner route: North Europe – Mediterranean – Far East

North Europe—Mediterranean—the Far East is the most popular common container liner route which calls Singapore, Shanghai, and Busan. There are 20 container liner services in this route which calls the three ports generating an average cumulative capacity of 312,096 TEUs per week. North Europe—Mediterranean—the Far East is also the most container liner routes in general. The second most popular liner trading route calling all three ports is the North America—Asia route, providing seven weekly container services with an average capacity of only 39,209 TEUs.

Container liner route: North America – Asia

Even though this route provides less cargo volume compared to the Europe—Asia route, this is an important trading route for both Shanghai and Busan. The United States is the largest trading for China with an export value of 418 Billion USD in 2019 (sources: UN Comtrade); and Busan is the largest port in the Far East region, making it the gateway between the two continents.

4.2 *Common Liner Services: Busan—Singapore (Exclusive)*

Container liner route: Intra Asia - NE Asia – SE Asia

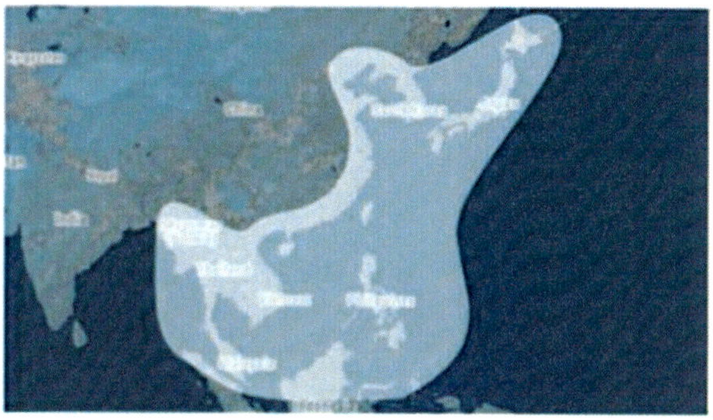

Only six container liner services are calling Busan and Singapore, which does not call Shanghai, implying Busan and Singapore do not depend on each other exclusively. This is reflected in the lower port intra-dependency ratio of only 2.65% and 1.83% for Busan and Singapore on the trade between the two ports, respectively. Since both the ports are transhipment ports, they compete for the same volume of cargo/container in the region, limiting the exclusive trade between the two ports. The liner routes between Singapore and Busan (and not calling Shanghai) are limited within Asia. There are four liner services trading exclusively between Busan and Shanghai, providing a container capacity of 13,988 TEUs weekly. There are also two liner services between connecting the Far East to the Indian subcontinent calling Busan and Singapore. Overall, it is observed that no long trading routes connecting Singapore and Busan exclusively. All the longer routes connecting Singapore and Busan has Shanghai as a port call.

4.3 *Common Liner Services: Busan—Shanghai (Exclusive)*

Container liner route: North America – Asia

Busan and Shanghai share a large number of common liner services increasing their port intra-dependency on each other. The port intra-dependency ratio for Busan on Busan—Shanghai trade is 36.53% while for shanghai is 22.16%. This indicates that Busan depends on the common services between the two ports more than Shanghai. Two major liner service routes which call Busan and Shanghai (and not calling Singapore) include North America to Asia and Intra Asia to Oceania routes. There are 39 liner services between North America and Asia which exclusively calls Busan and Shanghai every week with a cumulative capacity of 259,793 TEUs.

Container liner route: Intra Asia - Oceania

Intra Asia to Ocean is another major route calling the two ports, with 34 container services and an average of 58,951 TEU cumulative capacity every week. Small container vessels are allocated in this route which results in lower volume capacity. It is also observed that all the container routes between the Far East to Europe calls Singapore; hence there are no liner services in Europe—Asia route exclusively calling Busan and Shanghai.

4.4 *Common Liner Services: Singapore—Shanghai (Exclusive)*

Container liner route: Intra Asia

Container liner route: North Europe – Mediterranean – Far East

Though Singapore is a transhipment port and Shanghai receives its major cargo from the vast Chinese hinterland, the port intra-dependency between Singapore and Shanghai are comparable. The intra-dependency ratio for Singapore is 24.41% while for Shanghai is 21.45%. This refers that though Singapore has a high volume of transhipment containers, it has well-diversified its cargo volume and does not depend on only Chinese cargo. There are 20 liner services in North Europe—Mediterranean—the Far East, providing port call services to Singapore and Shanghai (and not Busan) with a cumulative of 165,766 TEU capacity weekly. This indicates that there are a large number of Europe to Asia trade which does not call Busan port.

Additionally, there are also 20 Intra-Asia container liner services which connecting North-East Asia to South East Asia through Singapore and Shanghai.

It suggests that there are opportunities for Busan by identifying and developing strategic partnerships with these liner services to generate higher cargo volume.

4.5 Liner Services: Exclusive to Busan

The port intra-dependency of Busan on itself is only 11.34% indicating that the Port of Busan depends extensively on cargo volume from Shanghai and Singapore. It also refers that Busan does not have well-diversified liner route portfolio for which it has a lower self-dependency ratio. Though Busan is a transhipment port that does not generate its own cargo, such low self-dependency ratio for Busan makes it vulnerable since any risks (such as geopolitical, financial or even pandemic) affect either Singapore or Shanghai will have a high impact on Busan's throughput.

Container liner route: Intra Asia

There are also no major liner routes which call Busan exclusively, that it, only calling Busan and not Shanghai or Singapore. It has a good liner network within Asia, providing 104 services exclusively to Busan with an average capacity of only 82,337 TEU weekly. It also suggests that, through Busan has a good network of services within Asia, the port still has to improve its connectivity beyond Asia to compete with the adjacent ports.

4.6 Liner Services Exclusive to Singapore

Unlike Busan port, Singapore being a transhipment port can generate a high volume of cargo on it own for which its self-dependency ratio is as high as 39.62%. It also indicates that Singapore relies on Shanghai and Busan for almost 60% of the TEU capacity (sharing common liner services) and it can generate about 40% of the volumes capacity from its exclusive liner services which do have Shanghai and

Busan has port calls. The high self-dependency might be attributed due to two reasons: (1) strategic location of Singapore which makes it the hub of transhipment for not only South Asia but also links between Africa and Oceania; and (2) even though there are other ports in Malaysia, Indonesia and Philippines location in the same region of Malacca Strait, due to improved operational efficiency, value-added services offered and well established strategic partners, Singapore is able to attract container liner.

Container liner route: Intra Asia – Africa – Oceania Services

Singapore is the central transhipment hub for Africa—Oceania, and Intra—Asia routes. There are 154 liner services in these routes providing a cumulative capacity of 321,194 TEUs per week. This not only provides a significant cargo volume to Singapore over the adjacent ports but also helps in diversification and risk management.

Container liner route: North America – Asia

There are also 16 liner services in North America—Asia route which exclusively calls Singapore (and do not call either Shanghai or Busan port) with a TEU capacity of 64,864 units weekly. Though it is a small volume of trade, it is unique as it strategically connects South Asia and Africa to North America. Hence Singapore not only stays futuristic, that is, connecting the production and labour hub to consumer hub, but also hedges itself for the US-China trade war diversifying its risks.

4.7 Liner Services Exclusive to Shanghai

Container liner route: Intra Asia – Oceania Services (including domestic)

Contrary to Singapore and Busan, Singapore is not a transhipment port, rather services as a major exporting port for China servicing its colossal hinterland. This also helps Shanghai to create its own cargo volume rather than depending on adjacent ports. So, the self-dependency ratio for Shanghai is relatively stable at 26.37%. There are 134 container liner services which call Shanghai (and do not call Busan and Singapore) in Intra Asia and Oceania route (including domestic routes). These services provide a cumulatively 191,905 TEUs on a weekly basis. There are also 14 exclusive liner services between North America and Asia with a 95,258 TEU capacity weekly.

Table 4 Port intra-dependency ratio overview

Port intra-dependency ratio	Busan	Singapore	Shanghai
Busan	11.34%	–	–
Singapore	–	39.62%	–
Shanghai	–	–	26.37%
Busan—Singapore	2.65%	1.83%	–
Busan—Shanghai	36.53%	–	22.16%
Singapore—Shanghai	–	24.41%	21.45%
Busan—Singapore—Shanghai	49.48%	34.15%	30.01%
Total	100.00%	100.00%	100.00%
Variance of the Dependency Ratio	4.73%	2.78%	0.16%

Container liner route: North America – Asia

These exclusive North America liner services do not call Busan, even though it is located at the intersection of both the continents. It also suggests that there is a high volume of trade between North America and China for which shipping lines can provide dedicated direct services.

4.8 Overview of the Port Intra-Dependency Ratio

Table 4 presents a summary of the port intra-dependency ratio for all the scenarios explained earlier. The results suggest that, amongst the three ports, Singapore has the highest self-dependency ratio, followed by Shanghai, and Busan has the lowest self-dependency ratio. On the contrary, Busan has the highest port intra-dependency ratio on the common liners between Busan, Singapore and Shanghai, indicating that Busan depends heavily on the co-operation between the three ports.

If we compare between Busan and Singapore since both are transhipment ports, it is observed that Busan has the highest dependency on Busan-Singapore-Shanghai liner services, followed by Busan-Shanghai liner services and then on itself. On the other side, Singapore has the highest dependency on itself, followed by Busan-Singapore-Shanghai and then Singapore-Shanghai liner services. This high dependency on a few adjacent ports can increase Port of Busan's risks, especially for the transhipment cargo volume. If Busan aims to become the transhipment hub in the Far East, it needs to reduce its dependency on Busan-Singapore-Shanghai services (say from 50% currently to less than 35% as that of Singapore) and needs to increase its self-dependency from about 11% currently to nearly 40% (similar to Singapore).

5 Case Study and Recommendations

After understanding the port intra-dependency of major ports in Asia—Busan, Singapore and Shanghai, this section focuses on Busan port as a case study to present the recent development of the port and addresses several challenges and opportunities which could support port authority and policymakers to build a strategic partnership with other countries, organizations and liner companies. It will help Busan port in mainly two folds: (1) to improve container throughput for meeting the growth of the port, and (2) to diversify the liner network connectivity for the port reducing the container throughput risks associated with the transhipment volumes.

5.1 Busan Container Terminal Development

The opening of the Busan port goes back to 1876, and the container port operation started since the 1980s. Since the 1990s, the Busan port has been top 10 largest[9] container port in the world.[10] Busan New Port's construction started in 1995, and until 2020, six out of ten planned terminals are operational. The satellite image of the old and new Busan port is presented in Fig. 5. The highlighted section on the right side of the image indicates the old terminals, while the left side presents the new container terminals.

There are nine dedicated container terminals in old Busan port; two midsize to Panamax container capacity, five terminals with Very Large Container Ships (VLCS) capacity, and two terminals capable of accommodating Ultra Large Container Ships (ULCS). The names of each terminal at the old Busan port is presented in Table 5. The old terminals do not have well-established railway connections. It can handle a total capacity of 6,677,000 TEU annually.

[9]The largest here refers to the annual throughput (TEU) of the ports.

[10]The following link presents a racing bar chart of top 20 container ports between 1995 and 2018: https://youtu.be/z2jAbU-UYxM.

Fig. 5 Old and New container terminal of Busan port. Source: Alphaliner

Table 5 Old Busan container port

	Terminals	Maximum Container size
1	U-AM Terminal	Midsized to Panamax
2	Hanjin Gamcheon Terminal	Midsized to Panamax
3	KBCT—Korea Express Busan Container Terminal	VLCS
4	HKT Hutchison Busan Container Terminal	VLCS
5	Busan International Container Terminal	ULCC
6	Hanjin Gamman Terminal	ULCC
7	Global Gamman Terminal	ULCC
8	Dongbu Pusan Container Terminal	ULCC
9	Hutchison Gamman Container Terminal	ULCC

Unlike the old container terminals, the new terminals are modern, capable of accommodating ULCS and are all connected with rail networks. The six operating container terminals have a cumulative capacity of handling 13.7 million TEUs annually. Table 6 presents the capacity of each terminal.

The container throughput of the new port has significantly increased vis-à-vis the old port since 2011. Currently, almost 68% of the total throughput is handled by the new port (see Fig. 6 for understanding the annual throughput trend for the two ports). In 2019, Busan port catered to almost 75% of the total throughput of the Republic of Korea.

Even though the total annual capacity of Busan port is 20.377 million TEUs, the port had a throughput of 21.992 million TEUs in 2019, indicating that some terminals had handled container above their capacity. Out of the entire throughput, 11.638 million TEUs (52.9%) were transhipment volume, 5.192 million TEUs (23.6%) were import volumes, and 5.163 million TEUs (23.5%) were export volumes. It is also observed that the local (export and import to the country) container

Table 6 New Busan container port

	Terminals	TEU Capacity
1	BNP 1-1 (PNIT—Pusan Newport International Terminal)	2,200,000
2	BNP 1-2	2,000,000
3	BNP 2-1b (HJNC—Hanjin New Port Terminal)	2,500,000
4	BNP 2-1a (PNC—Pusan New Port Terminal)	2,000,000
5	BNP 2-3—BNCT (Busan New Container Terminal)	2,500,000
6	BNP 2-2 (PSA—HPNT)	2,500,000
7	Busan New Port—2020 expansion A (generic name)	–
8	Busan New Port—2020 expansion B (generic name)	–
9	Busan New Port—2020 expansion B (generic name)	–
10	Busan West Container Terminal (3.5 B—Planned)	–
	Total	13,700,000

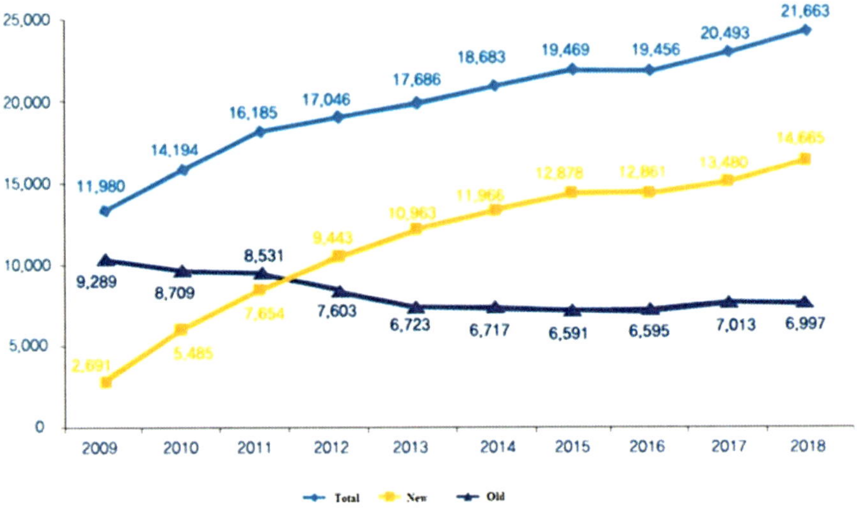

Fig. 6 Throughput trend: New vs. Old port. Source: Port of Busan Container Statistics 2019

volumes have been declining (from 55.4% in 2010 to 47.1% in 2019) and the transhipment volumes have been increasing (from 44.3% in 2010 to 52.9% in 2019). This indicates that the Busan port is heading from a traditional export/import port to a transhipment port. It also handles about 95% of the entire transhipment volume of the Republic of Korea.

5.2 Challenges and Opportunities for Port of Busan

Port of Busan faces fierce competition for the neighbouring Chinese ports. In 2001, Busan was the 3rd largest container port in the world, but it has dropped down to the 6th position in 2019. Port of Qingdao is marginally lower than Busan in 2019 with a high annual growth rate. Seven out of the top 10 container ports are located in China which attracts large volumes of cargo and limits the growth of Busan port. Additionally, Busan port has a high dependency ratio on common liner services calling Singapore and Shanghai port (as explained earlier in Sect. 4.8). Since over 50% of the throughput for Busan is transhipment cargo, this high dependency ratio for Busan on neighbouring ports can be detrimental for the port in both the short and long term.

In the short term, geopolitical tensions such as US-China trade war, pandemic, or any events and accidents can disrupt the connectivity with the adjacent ports and magnify its impact of Busan port. In the long term, the neighbouring Chinese ports, with their impressive growth rate, can take Busan's transhipment volume. Even though these ports serve the large hinterland of China and are primarily used for exports and imports, their activities cannot be ignored by the neighbouring ports. Furthermore, China is currently the largest manufacturing hub in the world, and Busan has strong liner connectivity with Chinese ports. However, as the economy of China is developing and the increasing labour cost can shift the production centres to other Asia countries like Vietnam, Thailand, Indonesia or even to Africa. Busan needs to develop strategic partnerships with other countries and diversify its liner network. This will not only decrease the dependency of Busan port on other regional ports but also improves container throughput resiliency for Busan under uncertainties.

Although Busan is the 6th largest port in the world, it is still growing continuously with an average rate of about 6.4% over the last 10 years. This nevertheless needs expansion to accommodate the growing demand. As there is a plan to establish four new container terminals in the New Busan port which would add about eight to ten million TEU capacity to the port, Busan needs to find additional volumes for this new capacity. Since the world GDP has slowed down, especially during the ongoing COVID-19 pandemic, the additional volumes can only be achieved by attracting new trading partners/markets and not from the existing markets.

Despite several challenges for the Port of Busan, there lie some interesting opportunities. The primary advantage of Busan port is its geographical location, that is, it is located in the junction of Asia (the manufacturing hub of the world because of the low labour cost) and North America (the consuming hub of the world because of the high GDP per capita) which is very similar to the strategic location of Singapore between Asia and Europe. All the liner services calling Asia/Far East to Europe has a port of call to Singapore. Unlike Singapore, there are 30 liner services (out of 159 services) calling between Asia (either Singapore or Shanghai or both) and North America which do not call Busan port. Busan port should develop strategic trade partners with shipping liner and countries to capture the trade between

Asia and North America and thereby increase the throughput. These 30 liner services have a cumulative capacity of 160,122 TEUs weekly. Further, as explained in Sect. 4.5, there are no major liner routes which are unique to the Port of Busan. All the major liner routes which are calling Busan are also calling Singapore and Shanghai. Busan should also think about developing competitive strategies such as game-theory concepts, network design and port competition by having unique major liner routes (vis-à-vis Singapore and Shanghai) to increase the market share and develop competitive advantages (Anderson et al., 2008; Cullinane et al., 2004; Yap et al., 2006).

Regional and Intra Asia trade is a way forward to improve the transhipment volume for Busan. It could utilize the downgrading (transhipment volumes) trend of the neighbouring ports such as Hong Kong as a case study (Wang & Slack, 2000) to monitor and reform its policies. This could help to understand the needs of several feeder ports and provide an efficient solution to improve the transhipment network for Busan port.

6 Conclusion

Currently, the international trade of commodities is facing turbulence and uncertainties due to the ongoing pandemic. The impose of port restrictions in China created an opportunity for Port of Busan to attract more container ships for loading, discharging and transhipment in the region, resulting in an increase in container throughput for few quarters in 2020. Macroeconomic uncertainties and risks can also increase the port market share for the players who have developed strategies and agility to encash the opportunities. This indicates that just being aware of the opportunity is not sufficient; rather, ports should be well prepared for availing these opportunities.

In this respect, this study explores the intra-dependency ratio between Busan, Singapore and Shanghai ports for containers using common and exclusive liner services from AIS data. It also provides a case study of several challenges and opportunities for the Port of Busan. Even though the maritime industry has been predominant and well-established industry, there has been limited usage of big data and technological advancement in decision making compared to other industries such as finance, pharmaceuticals or even aviation. The data related to the maritime industry, especially for ports, are not readily available. Hence the decisions related to ports are not always scientific; rather comprises of port managers, policymakers, amongst others who use their experience and qualitative assessments for managerial and operations related decision and develop long term strategies for ports. Combining management experience, along with evidence from quantitative data of shipping lines and port connectivity (container lines and trading route related data) can result in robust decision making.

A major takeaway from this chapter is that exploring big data in maritime and ports can complement the managerial decision making to (1) develop risk management strategies for ports and (2) retain as well as increase the port market share. While there was no clear evidence of when and how ports could benefit from the usage of big data in maritime a decade ago, there has been a growing interest and practices to build a decision support system for global supply chain management (Jha et al., 2020) and so as for ports/infrastructure. This emphasizes that the usage of big data in maritime is an emerging topic, where ports need to invest more in research to develop competitive advantage vis-à-vis other ports.

Acknowledgements The author would like to thank AXS Marine for providing access to Alphaliner that offers extensive container trade information, and extending their support to understand the novel database, without which study would not have been possible.

References

Anderson, C. M., Park, Y.-A., Chang, Y.-T., Yang, C.-H., Lee, T.-W., & Luo, M. (2008). A game-theoretic analysis of competition among container port hubs: The case of Busan and Shanghai. *Maritime Policy & Management, 35*, 5–26.

Asgari, N., Farahani, R. Z., & Goh, M. (2013). Network design approach for hub ports-shipping companies competition and cooperation. *Transportation Research Part A: Policy and Practice, 48*, 1–18.

Brandenburger, A. M., & Nalebuff, B. J. 1995. *The right game: Use game theory to shape strategy*, Harvard Business Review Chicago.

Cullinane, K., Fei, W. T., & Cullinane, S. (2004). Container terminal development in mainland China and its impact on the competitiveness of the Port of Hong Kong. *Transport Reviews, 24*, 33–56.

Desalvo, J. S. (1994). Measuring the direct impacts of a port. *Transportation Journal*, 33–42.

Hales, D., Lee Lam, J. S., & Chang, Y.-T. (2016). The balanced theory of port competitiveness. *Transportation Journal, 55*, 168–189.

Hoffmann, J., Asariotis, R., Assaf, M., & Benamara, H. (2018). *UNCTAD review of maritime transport 2018*.

Hume, D. (1752). Of the balance of trade. *The Gold Standard in Theory and History*, 330–341.

Jha, A. K., Agi, M. A., & Ngai, E. W. T. (2020). A note on big data analytics capability development in supply chain. *Decision Support Systems, 138*, 113382.

Jung, B.-M. (2011). Economic contribution of ports to the local economies in Korea. *The Asian Journal of Shipping and Logistics, 27*, 1–30.

Loichinger, E., Hammer, B., Prskawetz, A., Freiberger, M., & Sambt, J. (2017). Quantifying economic dependency. *European Journal of Population, 33*, 351–380.

Ma, S. (2020). *Economics of maritime business*. Routledge.

Muszyńska, M. M., & Rau, R. (2012). The old-age healthy dependency ratio in Europe. *Journal of Population Ageing, 5*, 151–162.

Notteboom, T. E., Parola, F., & Satta, G. (2019). The relationship between transhipment incidence and throughput volatility in North European and Mediterranean container ports. *Journal of Transport Geography, 74*, 371–381.

Park, J. S., & Seo, Y.-J. (2016). The impact of seaports on the regional economies in South Korea: Panel evidence from the augmented Solow model. *Transportation Research Part E: Logistics and Transportation Review, 85*, 107–119.

Sena, V., Bhaumik, S., Sengupta, A., & Demirbag, M. (2019). Big data and performance: what can management research tell us? *British Journal of Management, 30*, 219–228.

Shafaeddin, M. (1995). The impact of trade liberalization on export and GDP, growth in least developed countries. *UNCTAD Review, 1995*, 1–6.
Sinnathurai, V. (2013). An empirical study on the nexus of poverty, GDP growth, dependency ratio and employment in developing countries. *Journal of Competitiveness*.
Sloman, J., Garratt, D., & Guest, J. (2018). *Economics*. Pearson.
Song, D.-W. (2002). Regional container port competition and co-operation: the case of Hong Kong and South China. *Journal of Transport Geography, 10*, 99–110.
Tan, K. H., Ji, G., Lim, C. P., & Tseng, M.-L. (2017). Using big data to make better decisions in the digital economy. *International Journal of Production Research, 55*, 4998–5000.
Wang, J. J., & Slack, B. (2000). The evolution of a regional container port system: The Pearl River Delta. *Journal of Transport Geography, 8*, 263–275.
Wang, K., Ng, A. K., Lam, J. S. L., & Fu, X. (2012). Cooperation or competition? Factors and conditions affecting regional port governance in South China. *Maritime Economics & Logistics, 14*, 386–408.
Won, Y. J., Park, B.-C., Ju, H.-T., Kim, M.-S., & Hong, J. W. (2006). A hybrid approach for accurate application traffic identification. 2006 4th IEEE/IFIP Workshop on End-to-End Monitoring Techniques and Services. *IEEE*, 1–8.
Yap, W. Y., Lam, J. S. L., & Notteboom, T. (2006). Developments in container port competition in East Asia. *Transport Reviews, 26*, 167–188.

Part IV
Technology and New Business Model

Research Developments and Debates Regarding Maritime Autonomous Surface Ship: Status, Challenges and Perspectives

Tae-eun Kim and Jens-Uwe Schröder-Hinrichs

Abstract Recent developments in automation and telecommunication technologies have given rise to the prospect of autonomous ships, and inspired an exponential increase in research and regulatory discussions related to Maritime Autonomous Surface Ships (MASS) development and operation in the maritime industry. Propelled by several large MASS projects and trials in recent years, the area is evolving rapidly from a distant dream to a near reality by virtue of substantial interests and R&D investments. New vessel operational concepts based on remote-control and autonomous vessel functions are projected to increase efficiency and reliability, reduce operating expenditures (OPEX) and greenhouse gas (GHG) emissions, while having great potentials to contribute towards a range of safety and economic benefits to the ship-owners and the maritime industry in general. This paper provides a review of contemporary research studies relating to autonomous ship operations with the aim to contribute to current discussions in relation to two aspects: (a) it synthesizes the current research discussions and debates regarding MASS, and (b) it explores the possible barriers and hurdles associated with MASS operation. The paper concludes with suggestions for future research directions.

1 Introduction

We are currently experiencing a period of time with massive developments in areas related to digitalization and automation which is coined as the fourth industrial revolution (Schwab, 2016). The introduction of new technologies has a disruptive impact on all transportation sectors, of which maritime is no exception. From electronic Bills of Lading (B/L), paperless permits and reporting, to the maritime

T.-e. Kim (✉)
University of South-Eastern Norway, Vestfold, Norway
e-mail: tk@usn.no

J.-U. Schröder-Hinrichs
World Maritime University, Malmö, Sweden
e-mail: jus@wmu.se

B.-W. Ko, D.-W. Song (eds.), *New Maritime Business*, WMU Studies in Maritime Affairs 10, https://doi.org/10.1007/978-3-030-78957-2_10

single window concept, e-navigation, blockchain and Internet of Things (IoT) applications, remote-control and autonomous ship functions – the maritime industry is going through a phase of digital transformation. In particular, the development with regards to autonomous ship technology has generated a steep interest from scholars, media, maritime regulators, classification societies and labor organizations in recent years, mainly focusing on questions related to how and when it will unfold its full potential and what the implications will be for the maritime sector.

The field of MASS research and development is progressing rapidly, evidenced by several MASS projects (e.g., Yara Birkeland, AUTOSHIP, ReVolt, MUNIN, AAWA, SSAP, SVAN) and ship trials (e.g., Iris Leader, Falco). The proponents of MASS focus on benefits resulting from increased safety levels on board through relieving humans from unsafe and repetitive tasks, but also highlight the potentials for improvement in the economic and environmental performance of ships due to a proportionally higher cargo capacity, reduced OPEX in manning, reduced energy consumption and GHG emissions through innovative ship design, as well as reduced risks associated with human-related factors on board (e.g., fatigue, human errors, miscommunication) (Komianos, 2018; Kretschmann et al., 2017; Porathe et al., 2018; Pribyl & Weigel, 2018). Other benefits linked to MASS introduction could be positive effects in mitigating the shortage of seafarers, increasing the attractiveness of seafaring professions, as well as addressing a number of concerns related to seafarer welfare and gender imbalance issues in the maritime industry (Kim et al., 2019). However, operating commercial vessels autonomously also implies that the routines of ship operations, risk profiles, the shipboard arrangements, responsibilities and accountabilities will be different compared to conventional ship operation as of today (Kim & Mallam, 2020). This could result in significant changes of the global shipping business models, maritime education and training, public maritime policy and a different working environment for the 1.6 million seafarers around the world.

Advanced technologies and increasing automation are important prerequisites to enable the shipping industry continuing as the leading mode of transport (UNCTAD, 2020). To efficiently cover the advancing technologies by the international maritime regulatory framework, IMO has embarked on a Regulatory Scoping Exercise (RSE) for the use of MASS during its 98th session of the Maritime Safety Committee (MSC) (MSC98/23, 2017). The RSE involves two steps. The first step is an assessment of the applicability of the existing IMO instruments (e.g., SOLAS, COLREG, STCW 1978 as amended, STCW-F, Load Lines, SAR, Tonnage Convention, SPACE STP, STP) to ships with varying degrees of autonomous operation (IMO, 2018a). The second step is an analysis how MASS operations can be addressed, considering human, technological and operational factors (IMO, 2018a). This initiative is fully supported by a large number of IMO Member States and can be seen as an example for the proactive approach that IMO has adopted for maritime safety and marine environmental protection from ship-sourced pollution by balancing the potential benefits of innovative technologies against its operational implications for the wider maritime and marine environment. While the advantages of MASS seem to be clear to a larger community from a technical point of view, more divided views seem to exist on the timelines involved for the implementation of MASS in the maritime sector (WMU, 2019a).

In light of the above, this paper seeks to review relevant research studies relating to autonomous ship operations to round up the current research findings and debates regarding MASS, and to explore the possible barriers and hurdles associated with MASS operation.

2 Maritime Autonomous Surface Ship (MASS): Levels of Autonomy

Before we start the discussion regarding the specific factors impacting the implementation and uptake of MASS in the maritime sector, some very brief background information should be provided. In very general terms, automation has been defined as "*the use of control systems and information technologies to reduce the need for human work in the production of goods and services*" (Chialastri, 2012, p. 79). Hancock viewed automation systems as "*those designed to accomplish a specific set of largely deterministic steps, most often in a repeating pattern, in order to achieve one of an envisaged and limited set of pre-defined outcomes.*" (Hancock, 2019, p. 481), which is different from autonomous systems that are defined as "*...permanently change their functional capacities as a result of the input of operational and contextual information.*" (Hancock, 2019, p. 481).

Automation of systems in ships has been an ongoing process since the first industrial revolution started with the introduction of the steam engine and the automatic control systems that were required to effectively and safely operate these engines. Shipping, like any other industry sector, has been transformed since those early days of automation quite significantly and the automation systems of today are very much different from the days of the early attempts to automate systems in industrial devices.

IMO initially started discussions regarding automation of ships to be covered by maritime safety regulations as early as 1964 during the eighth session of the Maritime Safety Committee (MSC) (WMU, 2019b). This was a first attempt to identify trends and developments in automation and technology that would require a regulatory approach to ensure that any technical development would not compromise the safety performance of ships. When the initial discussions related to autonomous ship technologies reached a certain level of technical feasibility, a group of Member States of IMO was making inquiries if it was about time to discuss these technologies in the relevant Committees of the Organization. Following a proposal submitted by Denmark, Estonia, Finland, Japan, Netherlands, Norway, Republic of Korea, United Kingdom and United States in 2017 (MSC98/23, 2017), MSC agreed to initiate a RSE to review regulatory implications arising from MASS and to plan for required future activities on a regulatory level (MSC98/23, 2017). Several States have begun to initiate MASS guidelines and regulatory frameworks, and a number of

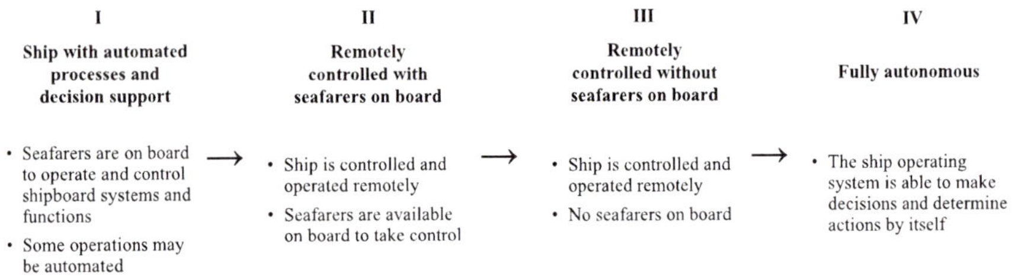

Fig. 1 Ship autonomy levels based on the definitions of IMO (2018b)

organizations (e.g., Lloyd's Register, Rolls-Royce, Bureau Veritas, Norwegian Forum for Autonomous Ships (NFAS), UK Marine Industries Alliance, Ramboll) have proposed definitions for concepts and levels of ship autonomy (MSC99/5/6, 2018). To facilitate the process of the RSE, IMO has defined a MASS as *"a ship which, to a varying degree, can operate independent of human interaction"*(IMO, 2018b) and has proposed four levels of autonomy of MASS, as illustrated in the following Fig. 1: (1) Degree 1: ships with automated processes and decision support; (2) Degree 2: remotely controlled ships with seafarers on board; (3) Degree 3: remotely controlled ships without seafarers on board; (4) Degree 4: fully autonomous ships (IMO, 2018b). It was noted that MASS may have the possibility of operating at different degrees during a single sea voyage (Kim & Mallam, 2020).

In a recent study by WMU (2019a), six factors were highlighted as enabling factors of MASS, which include economic benefits, technological readiness, social acceptance, knowledge and skills, labor market dynamics, regulation and governance. In addition to these, the great potentials of MASS for safety, environmental and efficiency gains, as well as the market competition and industry need have also been the main motivations for MASS development in the maritime industry as suggested by MASS proponents. The key motivating and enabling factors of MASS have been summarized in the following Fig. 2.

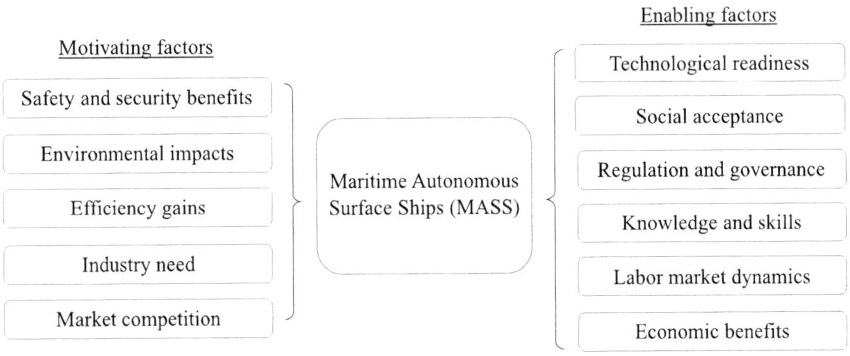

Fig. 2 Motivating and enabling factors of MASS

The remainder of this paper will review relevant publications with regards to these factors in order to provide for an overview of the current developments and the state-of-art in relation to MASS.

3 Research Development and Discussions Regarding MASS

While evidence of the potential benefits of MASS mount, there are still many unsolved problems and challenges to its development and widespread adoption. Those problems and challenges are not only associated with the technical capabilities (e.g., automatic navigation control, collision avoidance), but also economic viability, safety, security, legal, ethical, as well as management aspects of MASS operations. All of these aspects have been investigated—to varying degrees—in a number of MASS R&D activities and exploration endeavors.

3.1 MASS Technological Feasibility

Any new technology has to reach a certain technical readiness level before the market is prepared for an uptake of this technology. The review of current developments and the state-of-art in MASS should start with an evaluation of factors related to technical feasibility. This creates the basis out of which any of the other factors outlined before evolves.

There are several key systems that enable the vital technical functions of a ship. To provide a comprehensive overview about the status of all of them in a short paper is certainly a challenge. We will therefore concentrate on two key systems - navigation and communication and only briefly touch upon other relevant systems and aspects.

Embracing the use of automation technologies in ship operations is not new for merchant vessels. The ship automatic pilot (autopilot) control system installed onboard of the modern vessels today is able to follow the manually input course and maintain the set heading automatically (Burns, 2001, see section 1.3.4). Autopilot has been evolved over the years to improve course-keeping performance by considering the ship dynamics and environmental impact on the course of a ship (Roberts et al., 2003; Velagic et al., 2003). The use of autopilot relieves humans from constant manual course changes and alterations, minimizes rudder movements which subsequently reduce fuel consumption and OPEX. However, autopilot is not recommended when navigating in rough weather conditions, or narrow, confined or congested water, as it is not sufficient enough to observe and detect obstacles as well as to turn the vessel accordingly to avoid collisions. Human intervention is needed to constantly monitor the environment and to manually maneuver the ship in certain areas and sea condition (Raunek, 2020). The autopilot as described above, is a key system on current ships. It is clear, that its functionality is of even greater importance on MASS where it requires further development to navigate the ship without constant human supervision. It therefore requires a system for automatic navigation control but also collision avoidance function that functions in all the dynamic sea conditions and traffic situations (Lyu & Yin, 2018). This requires designing a control process to constantly sense, avoid, adjust the control of the vessel based on the real-time feedback generated through sensors, cameras, GPS, radar, AIS as well as other monitoring equipment. Independent from the goal of MASS, steady technological development in this area would certainly be a good aid to navigation of ordinary ships, especially during poor visibility due to darkness and fog.

There is a wide range of collision avoidance control algorithms available as of now that follows the International Regulations for Preventing Collisions at Sea (COLREGS). However, they still face technical challenges in generating safe and optimal paths in complex navigational scenarios (Johansen et al., 2016). In this respect, a considerable amount of research has been conducted looking into the collision avoidance aspect of MASS (Abilio Ramos et al., 2019; Huang et al., 2020; Johansen et al., 2016; Johansen, Perez, & Cristofaro, 2016; Lee et al., 2015; Statheros et al., 2008; Zhao & Roh, 2019). According to Zaccone and Martelli (2020), ship autonomous collision avoidance systems include three key elements: an obstacle detection system (for identifying obstacles in vicinity and to provide situational awareness), a motion planning system (for generating a safe path based on the identified obstacles), and a motion control system (for maneuvering the ship based on the identified path) (Zaccone & Martelli, 2020). A large number of studies have been conducted to investigate into these elements and to explore the motion control for autonomous navigation (Abilio Ramos et al., 2019; Kim & Kim, 2018), path planning (Bitar et al., 2018; Chen et al., 2020; Hassani & Lande, 2018; Lyu & Yin, 2019), as well as large roll motion detection (Acanfora et al., 2018), intelligent rudder control (Larrazabal & Peñas, 2016), and ship docking (Shuai et al., 2019). To ensure that the autonomous ship operating system is able to generate safe and efficient trajectories in different maneuvering situations and in unfavorable weather conditions is a fundamental prerequisite for autonomous operation of ships.

When the technical design of ships is concerned, SOLAS is the key instrument to consider. MASS need to find ways to address the various technical specifications in the different SOLAS Chapters in their own specific way. An important aspect in this context is fire safety. This raises the questions how to design required technical systems able to prevent and defend MASS in the event of fire and other emergence situations (Wróbel et al., 2017). Furthermore, SOLAS requires every ship to provide assistance to vessels in distress in its vicinity. How to enable MASS for such assistance of other vessels operating close to a MASS and to accommodate survivors without crew members onboard that would normally be able to provide active support on site should also be considered (Dreyer & Oltedal, 2019).

For MASS, more than for any other ship, communication and data transfer is a key component, which focusses on arrangements for information sharing and communication system for both ship-to-ship and ship-to-shore communication (Dreyer & Oltedal, 2019; Kooij et al., 2018). This includes issues related to how fast and reliable the information generated from the sensors, cameras and other equipment can be transferred between the ship and the operators ashore to enable wireless monitoring and real-time ship handling (Hoyhtya et al., 2017), as well as how to enable ship-to-ship communication depending on the degree of autonomy to facilitate situation awareness and coordination in both normal and distress situations. The ongoing e-Navigation initiative from IMO (MSC85, 2009) to standardize the information between ship and shore could also facilitate future MASS operations. Very few studies to date have specifically explored the communication aspects of MASS. Research undertaken so far has reviewed the possibility of using 5G as a potential maritime communication service carrier (Ait Allal et al., 2020); whether state-of-art satellite communication systems are suited for safe operation of MASS (Namgung, 2019); as well as challenges for MASS in the area of information technology and telecommunication (Yanchin & Petrov, 2020). There is also a debate among experts whether or not the current bandwidth of the available satellites is sufficient to support the data communication needs of a larger fleet of MASS globally. Considering the potential consequences of communication outages for MASS, it would be important to extend the scholarly discussions and developments in this area and to review to which extent the currently available satellite infrastructure needs to be expanded, which would require substantial investments.

Another important aspect is the system integration. All the different systems outlined before have to be integrated in a way that they work together in a safe and reliable way. This is an enormous challenge, especially in complex operational situations. In case of a leak and related inrush of water, a damage, mechanical or communication system failure and other challenging circumstances, a MASS needs to have the ability to detect and evaluate the situation, to develop a response strategy and to implement appropriate actions for, e.g., repair and maintenance (Heffner & Rødseth, 2019). In addition to above mentioned issues, technological challenges also lay in vehicle health management system, automation of the propulsion system, cargo handling, system integration, as well as the onshore infrastructure development with regard to remote-control center and telecommunications infrastructure.

Finally, another important aspect not discussed so far in this paper, relates to propulsion systems of MASS. Combustion engines currently used on commercial vessels require constant maintenance and supervision by an engine crew and therefore set limits to autonomous operations on highest levels. Autonomous operations of engine rooms so far involve electric propulsion systems (e.g., Yara Birkeland). Such electric propulsion systems require regular exchange of the batteries (refer to the Zero Emissions Services (ZES) Initiative in the Netherlands) and therefore are an encouraging step providing initial options for inland waterways and short sea shipping. However, this does not provide a basis for ocean-going long-distance operation of MASS, yet.

3.2 Safety and Security Issues

The evolution of the technical feasibility of MASS creates the basis for this new type of ships and shipping operations. Considering that MASS are still at a very early stage of development, it is clear that most publications and research activities focus on the technical aspects of and potential solutions for MASS. However, as MASS concepts mature, other relevant factors for the uptake of MASS globally enter the scene and need to be addressed by the MASS promoting community. Apart from the technological aspects, identifying an appropriate MASS management structure, safety and security assurance plan, shipping governance framework, reskilling and upskilling framework are also essential issues that need to be addressed on the road to fully developed MASS. This requires an in-depth understanding regarding the nature and side effects of the technology needed, its operational process, capabilities, constraints, emerging risks and hazards, as well as human factors in parallel of technological development.

MASS will fundamentally change the operational concept of shipping operations. This also implies the emergence of new hazards, risk profiles and security concerns, which calls for new measures to mitigate or eliminate them. Researchers have noted that automation does not simply displace human tasks with machine tasks, it also creates new error pathways and could potentially bring a set of unanticipated risks to the shipping system (Lützhöft & Dekker, 2002; Porathe et al., 2018; Rigaud et al., 2012). Introducing new technologies in the maritime industry has always led to new accident types and failure pathways as a result of unanticipated side effects of these new technologies (Perrow, 2011). AIS, ECDIS and RADAR, to name a few examples, have been introduced to aid navigation, but also created new types of accidents (Schröder-Hinrichs et al., 2016).

Many researchers have studied the hazards and risks associated with MASS operations (Chaal et al., 2020; Chang et al., 2021; Valdez Banda et al., 2019; Ventikos et al., 2020). Various hazard analysis techniques have been applied to develop a better understanding of the potential risks of MASS operation. Applications include Fault Tree Analysis (FTA) and Failure Modes and Effects Analysis (FMEA) to understand risks associated with MASS (Ait Allal et al., 2017; Chang et al., 2021). Furthermore, systemic hazard analysis techniques, such as the System Theoretic Process Analysis (STPA) have been applied in several studies (e.g., Chaal et al., 2020; Ventikos et al., 2020). Risk factors were identified and possible accident scenarios evaluated, so that potential risk control options could be identified to address those risks and hazards during the system design stage or during normal operations before any damage may occur. Different hazard analysis techniques are based on different causality assumptions (Leveson, 2011). The mix of different techniques has helped to provide a comprehensive picture of the relevant hazards associated with MASS operation. Chang et al. (2021) synthesized the literature findings into six main hazard categories, including (1) system failure (e.g., system malfunction, communication breakdown), (2) human error (e.g., errors in system design, programing, remote control), (3) equipment failure (e.g., failures of sensor, electricity, IT structure, communication system), (4) cyberattacks (e.g., malware, information theft, or other cyberattacks for financial profits, reputation damage or military purposes), (5) interaction with physical environment (e.g., poor weather condition, area of icing), (6) interaction with manned vessels and detection of objects (e.g., risk of collision with other vessels, failure to detect semi-submerged and small objects). Apart from identifying these main hazard categories, their study has further revealed that the "interaction with manned vessels and detection of objects" (in particular the failure of detecting semi-submerged objects) and cyberattacks are ranked as the two top hazards in MASS operations, followed by human error-related hazards and equipment failures.

Adopting new Information and Communications Technologies (ICT) enabled operation improves efficiency and performance. However, this also leads to increased cyber risks in many industries (Kavallieratos et al., 2018). Cyber-attacks and general security issues of MASS, communication systems and data security using blockchain technology have also been considered in several studies (Kavallieratos et al., 2020; Tam & Jones, 2018; Valdez Banda et al., 2019). Results have shown that AIS, ECDIS and the GMDSS as well as human machine interfaces and the navigation systems may face higher cyber risks (Kavallieratos et al., 2018). With regards to physical security issues of MASS, it can be envisaged that unmanned and autonomous ship operations may involve a higher likelihood of illegal boarding and/or robbery. Whilst the consequences of pirate seizure might be reduced when the ship operates without a crew, the security issues regarding communication and IT systems of MASS would be as important as other security measures (Honekamp, 2018).

3.3 *Technology Governance*

The prospect of MASS has not only triggered the discussions regarding its ethical, legal, social impacts, but also the regulatory changes and steps that need to be taken at a national and international level to fully utilize the benefits of those emerging technologies while preventing, controlling and/or mitigating potential negative consequences. "*Many of the barriers to enabling emerging technologies lie not in technology per se, but in technology governance*" (OECD, 2020).

David Collingridge (1982) has introduced the "Collingridge dilemma" in his book "the social control of technology", which is frequently cited in the discussion of emerging technology governance (e.g., 3D printing, Artificial Intelligence (AI), autonomous vehicles), but not yet in the context of autonomous ships. He describes that the social consequences of a technology cannot be fully anticipated at the early stage of its development. However, when the undesirable social effects are discovered, the technology has often already become a part of the economic and social structure, which makes it difficult to control it at this stage. This encompasses the dilemma of control. When change is still easy at an early stage of development, the need for it may not be fully anticipated. Conversely, when a need for change is identified at a later stage, that change becomes difficult, expensive or time-consuming, making it difficult or impossible to change. This phenomenon was coined as "Collingridge dilemma" and has become an essential point of reference in technology assessment discussions (Genus & Stirling, 2018; Liebert & Schmidt, 2010; Mayne, 1982).

One way to deal with the dilemma is to undertake an early assessment of the potential and unforeseen side effects of a technology about to be taken up in a technological development process (Buckley et al., 2017). The awareness for potential ethical issues and societal impact associated with its expected trajectory could steer the design of this technology and its control system to avoid those anticipated effects before entering into the market. However, uncertainty and ambiguities involved in technology development make the early assessment and rule making process rather complicated. To regulate ahead of time has always been challenging. Disparity or gap between a technological development and the development of legal mechanisms to govern this technology has been frequently described (Marchant, 2011; Moses, 2007).

If this gap is too large, when, i.e., the technological development outpaces the regulatory processes, the industry would work without sufficient legislative and ethical oversight of the technology under development which would not allow for effective governance of the sector. In such a situation, different priorities may exist other than, e.g., ensuring safety. It may also be difficult or impossible to initiate corrective actions or to impose sanctions for misuse and infringements. This pacing problem has received increasing scholarly attention in recent years and the question how the legal frameworks could keep up with the technological development has been discussed in a number of studies (Bennett Moses, 2011; Hagemann et al., 2018; Marchant et al., 2011). Embracing a proactive and flexible regulatory approach through "soft laws", which is generated through industry best practices and standards, responsible corporate behavior, stakeholder consultations etc., has been considered as an essential tool in filling the pacing gap (Hagemann et al., 2018; Huddleston, 2019; Marchant et al., 2011). The development of interim regulatory measures in member states, and the RSE initiative taken by IMO to review how MASS operations could be addressed in various IMO instruments can be regarded as an important and proactive effort in minimizing the pacing problems in the maritime industry in this respect. The potential benefit of proactive measures in addressing the MASS pacing problem is also shown in Fig. 3 below.

A number of recent papers have discussed the regulatory barriers and legal issues related to MASS and its operations (Bačkalov, 2020; Karlis, 2018; Li & Fung, 2019; Ringbom, 2019; Vojković & Milenković, 2020). The maritime industry is currently regulated by comprehensive conventions with clear lines of responsibility among the actors. Remotely-controlled and unmanned ship designs and operational concepts are not yet covered by current maritime regulatory frameworks. There are more than 61,000 conventional ships which carry more than 80% of world trade today on the global oceans (Allianz, 2020; IHS, 2019; UNCTAD, 2020). It can be projected that we will approach a reality where MASS at varying degree of autonomy and conventionally crewed ships operate at the same time in the same sea area. This

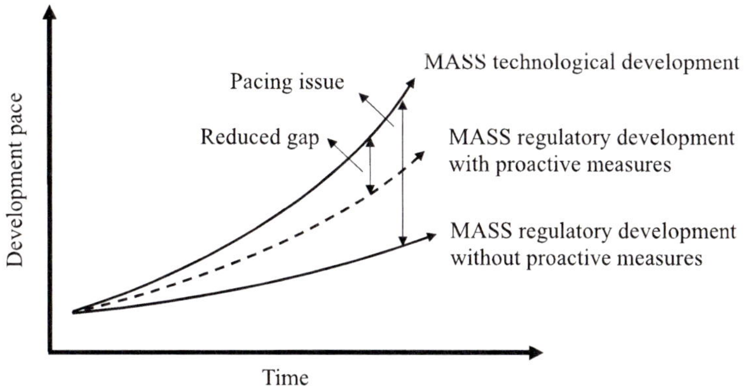

Fig. 3 Importance of proactive measures for addressing MASS pacing problem

creates challenges for the global maritime governance and the international legal instruments. No specific ideas have been voiced how MASS would be certified, how their seaworthiness is ensured, how MASS could be insured, whether (or how) remote operators can and should be considered as seafarers or not, under which jurisdiction and regulatory regimes remote-control center and their operations will fall, who should be liable in the event of an accident, and whether developing a new instrument for MASS is more efficient and favorable than amending existing instruments, etc. These are only a few selected issues in relation to technology governance. The key international instruments for global maritime transportation (e.g., UNCLOS, SOLAS 1974, COLREG 1972, STCW 1978 as amended) are based on the assumption that a ship is controlled by a human operator onboard a ship. Article 94 of the United Nations Convention on the Law of the Sea (UNCLOS) defines the duties of a flag State and requires the flag states to take measures to ensure that each ship is operated by a master and officers who possess appropriate qualifications (UNCLOS, 1994).

Adding automated functions onboard ships to assist crews during their daily operations may not pose significant challenges to the current legal system. However, when the degrees of autonomy increase (which involves remote-control operators, unmanned bridge, artificial intelligence, machine learning-based decision-making system) it may be challenging to govern these new operational modes with the existing rules and liability frameworks. A remotely controlled MASS may be covered by a similar liability regime like a conventional vessel. The main difference is the transfer of tasks from the shipboard operators to the remote-control operators. Defining the role and responsibility of remote-control operators would be essential for determining the liability of a remote-control center (Lee, 2016). However, at autonomy degree 4 of MASS operation, the shipboard duties are completely replaced by computer based operating systems and shipowners may have only limited (or no) control of their own vessels. This would create uncertainties in relation to how liability should be regulated in case of damage to the vessel, cargo, shore infrastructure, etc. The liability considerations for fully autonomous vessels would be far more complex than human controlled or remotely controlled MASS (Lee, 2016).

However, it should be noted that the challenges in regulating emerging automation technologies are not new for a variety of industries in the world. Similar technology regulatory and governance challenges have been observed and discussed in many industrial sectors e.g., autonomous vehicles on roads and the healthcare sector (Gasser & Almeida, 2017; Reddy et al., 2020; Stilgoe, 2018). While the debate about regulatory frameworks is still ongoing (Daly et al., 2019; Salathé et al., 2018), many countries have participated in this debate related to emerging technology governance with the objective to ensure that there is an even level playing field and to raise the bar for responsible operations and developments without delaying the pace of innovation (Daly et al., 2019).

Giving the early stage of MASS development, instead of a rigid prescriptive regulatory framework, a governance model with key ethical principles and norms to provide for an axiological basis for the technological development and operation could be an innovative step forward, so that the legal system could respond to concerns and ethical questions regarding remote-control and autonomous operation while retaining flexibility as new opportunities emerge.

3.4 Economic Viability

While technical feasibility and regulations are key enabling factors for MASS, this technology will not experience an uptake if its economic viability and commercial attractiveness cannot be addressed. This is a fundamental prerequisite for shipowners to invest to this new operational concept. There is a divided view about the economic viability of autonomous ship in the maritime community. The ship construction costs of MASS, the life cycle costs, training cost for personnel on board, remote control infrastructure cost, maintenance cost, as well as other hidden costs of MASS cannot be clearly determined at the moment as there are very few academic studies only that have been conducted on the economic viability of MASS.

Kretschmann et al. (2017) explored the economic benefits of MASS by comparing the Net Present Value (NPV) and Required Freight Rate (RFR) of an autonomous bulker with a conventional one under three scenarios over a 25-year period. NPV (i.e., all net earnings the vessel is expected to generate during its commercial life) has to be positive in order to make an investment worthwhile. RFR (i.e., the amount to be charged in order to break even) should be lower in comparison for better economic performance. They have reviewed three potential scenarios. According to their analysis, in a reduced crew (with new shore and port services) scenario, MASS do not have significant economic advantages over a conventional vessel. This is indicated by the results of an RFR of an autonomous bulk carrier delivering a zero NPV which is 0.4% lower in comparison to the conventional vessel's RFR (Kretschmann et al., 2017). However, it is likely that the reduced crew scenario will be attractive for high-cost countries such as Norway (Akbar et al.,

2020). When considering a scenario with reduced crew and improved fuel efficiency, the required freight rate of the autonomous bulker which produces a zero NPV is 3.4% lower in comparison to the conventional vessel's RFR, which indicates that the autonomous vessel has better economic profitability (Kretschmann et al., 2017). In the last scenario with reduced crew, improved fuel efficiency and with the use of MDO as the main fuel instead of HFO, the autonomous solution would not be financially viable as the expected present value of costs of the autonomous ship is significantly higher than the conventional one (Kretschmann et al., 2017). We discussed this paper in all details, as it is one of the few studies in this area and also indicates that MASS does not immediately provide strong and significant economic benefits in comparison to conventional ships.

Another dedicated economic analysis from Akbar et al. (2020) has found that the introduction of autonomous daughter ships could decrease the total operating costs by 11% on average in comparison to conventional ship. The reduced time charter cost contributes to most of the cost savings (Akbar et al., 2020). Despite the high initial investment, reducing crew cost, coping with GHG measures and using free available space for more cargos would increase the ship earning capacity and efficiency while reducing the OPEX.

With regards to the question if the concept of maritime autonomous ship operation is just a hype or if we are indeed at the dawn of a new era of the maritime transportation, the following comments can be offered.

Similar to other industrial sectors (e.g., aviation, vehicle), shipping will also gradually move towards a more autonomous future as the beneficiaries of technological advancement, increased digital connectivity and artificial intelligence become more obvious and real. It is therefore not a question of *if* MASS will happen but *when*. However, it will rely on a broad array of factors before its full implementation and widespread adoption. These factors do not only include the technological readiness, but also economic incentives, emission regulations compliance, and workable financial solutions for the shipowners to consider adopting a new vessel operational concept.

The availability of MASS in the market may also lead to a two-tier system with conventional commercial ships being discounted heavily against MASS. How the industry is going to accept and operate these new ships under current and near-future shipping market would be worth of further exploration in other academic studies.

3.5 Human Element and Social Implications

As highlighted before, while the majority of studies published in relation to MASS has focused on technological issues, technical feasibility is only one aspect to be considered when discussing the uptake of MASS. Another important aspect relates to the human element aspects of MASS. We can briefly divide the meaning of "human element" into two parts: human resources and human factors. The former would refer to the people and all the players associated with ship operation, whilst the latter part is concerned with using our understanding of human behaviors, physical and psychological characteristics to improve design of the system for better, safer and more efficient human use.

One of the key benefits suggested in the context of the implementation of MASS is its potential to decrease negative impacts of human factors on safety, which could in turn improve the safety performance of the maritime industry. However, scholars have been skeptical about this proposed benefit of digitalization and automation and argue that human factors will still remain a critical issue for safety (Ahvenjärvi, 2016; Dreyer & Oltedal, 2019; Kim & Mallam, 2020; Lützhöft & Dekker, 2002; Porathe et al., 2018; Wróbel et al., 2018).

Automation could displace human operators from labor-intensive, routine, repetitive and error-prone tasks. However, the increased use of automated systems also implies distancing human operators from the direct system control and operations. This has often led to several commonly recognized human factor issues, such as out-of-the-loop syndrome, failure to develop an accurate level of situational awareness, information overload, boredom, skill degradation, 'look but do not see' problem, complacency, technology over-reliance (Dhami & Grabowski, 2011; Endsley, 2018; Pazouki et al., 2018; Wahlström et al., 2015), which may consequently lead to different, but equally severe risks in more automated ship operations. In order to experience the full potential and benefits of automated system, system designers should take human factors into full consideration during system design and the development process. Both, the automated system and human operators, have limitations. Human operators may not achieve the level of reliability that automated systems maintain. Automated systems may not be able to handle unanticipated situations as resilient and adaptive as humans. Therefore, instead of taking a technology-centered approach, the system designers should consider placing the users, i.e., the onboard crew members or remote-control operators, at the center of MASS system design to enable them to fully utilize the potentials and capabilities of automation and also to control the limitations of automated systems.

Reason (1995, p. 85) noted that, "*Increased automation does not cure the human factors problem, it simply changes its nature*". In many scenarios, automation would lead to changes in the nature of tasks and task structures, which would demand a new skillset of the operators (Lee & Seppelt, 2009). Implementation and operation of MASS would redefine the roles and responsibilities of the shipboard officers and engineers (e.g., masters, chief officers, chief engineers and second engineers) and re-allocate their tasks and duties. Skillsets and competence requirements will differ from those required on conventional ships (Kim & Mallam, 2020; Sharma et al., 2019). Formulating a reskilling and upskilling framework is also an essential issue that need to be addressed on the road to MASS. However, there has been very limited research looking into the required competence framework for MASS operators under various degrees of autonomy. Among the limited empirical studies which focuses specifically on the impact of MASS on STCW competence frameworks, Sharma et al. (2019) have explored the impact of autonomy degree 2 operation on seafarer competence requirements with a specific focus on the potential changes that may be required among the Knowledge, Understanding and Proficiency

(KUPs) items for officers in charge of a navigational watch as listed in STCW Table A-II/1. The current STCW Table A-II/1 contains 19 competence themes, consisting of 66 KUPs which define the minimum competence standard for officers in charge of the navigational watch on ships of 500 gross tons and above (IMO, 2011). The preliminary results of this study have shown that many of the KUPs required today may become obsolete in the future and more emphasis has been added on competences related to, e.g., emergency management during autonomy degree 2 operation. Further KUPs should be added to the re-skilling framework and related efforts are required to effectively prepare seafarers for the shipping industry of the future (Sharma, Kim et al., 2019). Sharma and Kim (2021) have further commented that future officers should also be well equipped with appropriate digital competence, software engineering, cyber-attack defense skills, as well as engineering competence to be able to control and master the challenges created by diverse types of equipment and systems to be used onboard MASS. In terms of changes in relation to the non-technical skills, Kim and Mallam (2020) have investigated the applicability of the current leadership KUPs as well as the new leadership competence requirements that need be developed by shipboard officers and remote-control operators for crewed, un-crewed and remotely controlled MASS operating at autonomy degree 1 and 3. The authors present that the "knowledge and ability to apply decision-making techniques", "knowledge and ability to acquire, handle and comprehend a large amount of system information" were deemed to be the most critical non-technical skills to be developed by shipboard officers operating at autonomy degree 1 (Kim & Mallam, 2020). In contrast, for remote-control operators of unmanned ships the ability to obtain and maintain adequate situational awareness is perceived as the most critical competence requirement for autonomy degree 3 by the subject matter experts (Kim & Mallam, 2020).

Benefits of automation have been amply documented in research studies focusing on the transportation sector (Litman, 2020; National Research Council, 2014) and other industrial sectors (Brettel et al., 2014; Hofmann & Rüsch, 2017). However, automation also leads to a situation where certain tasks in a job profile will be handled by technology and could make jobs redundant. As the automation technology is able to take over further functions during ship operations, a controversial debate has started to discuss the implication of MASS in relation to human resources in the maritime industry. One issue in this debate is the implication of technological unemployment in the context of MASS. Many shipboard functions could be handled by technology and lead to reductions of crew. Several recent studies have discussed the impact of ship automation on employment (Kim et al., 2019; WMU, 2019a). The WMU report (WMU, 2019a) suggests that despite technological advancement with the potential to replace many roles onboard ships, there is still demand for humans in terms of maintenance, screening, surveillance and security controls for a foreseeable time. Low skill job profiles may be at greater risk due to the introduction of MASS, whereas high skill job profiles could be affected in a different way as they may be supported by technology and automation but not necessarily replaced (WMU, 2019a).

Seafarers have been the key on-scene decision makers, operators, and trouble shooters onboard of ships today, whose seamanship, adaptability, problem solving skills and sound judgment have enabled nearly all ships to arrive at their destinations without an accident. A small percentage of unsuccessfully performed operations leading to maritime accidents are inevitably present and have the potential to result in significant economic, reputational and legal consequences and societal impact (EMSA, 2019; Kim et al., 2016). The results from a study of Wróbel et al. (2017) suggest that the introduction of MASS could reduce the likelihood of navigational accidents, such as collisions and groundings, while the probability for non-navigational accidents caused by fire, explosions, flooding, structural failure would be greater for unmanned ships due to the absence of onboard crew members who could mitigate the consequences of an accident at an early stage of the accident development. The question if MASSs are safer than conventional ships has not yet been fully explored. The human element is a key element for maritime safety and the safety of navigation and therefore remains a key priority in the strategic directions of IMO. On the other hand, the industry is moving forward in changing the conditions under which the human element is manifested in shipping operations. Seafarers are the key enablers of global trade. Placing them at the center of MASS system development and adopting a more user-centered approach to automation system and interface design should be appropriately considered.

Another issue with regards to the impact of MASS on human resources is skill degradation. With increasing reliance on automated systems and a reduced need for direct maneuvering activities, skills may fade and degradation might occur. Lutzhoft, Hynnekleiv et al. (2019, p. 12) noted that skills should not only be updated in accordance with the demands of the new technology, but should also be protected from eroding so that they can be used in the event of system failure. Many studies exist on the subject of skill degradation with automation (Bainbridge, 1983; Form, 1987; Volz et al., 2016). However, very few academic studies to date have focused on skill degradation in the maritime industry. Future research in this area could look into the skill fade and degradation issue in an era of autonomous shipping and identify measures to combat and/or mitigate these effects.

4 Reflections and the Ways Forward

Building and introducing an autonomous ship to a local market/scene may be done by a single organization. However, adopting, accepting and operating of an autonomous oceangoing ship would require collaborative efforts of a large group of maritime stakeholders at a global level. A critical challenge for autonomous shipping is how the key maritime stakeholders would optimally balance collaboration and competition to support the successful transition to a new era of shipping.

The technology needed to ensure remote-control and autonomous shipping has not yet reached its zenith. Many uncertainties remain and many fundamental questions are yet to be answered. Are MASSs safer than (or as safe as) conventional ships? Are MASSs more profitable than conventional ships? How will the technical systems be designed to prevent and protect MASS in the event of fire and other emergency situations? How to enable MASS for assisting other vessels in the vicinity of a MASS and to accommodate the survivors without onboard crew members' assistance? How to enable wireless monitoring and real-time ship handling? Who should be legally liable in the event of an accident? What is the need and reason for the maritime industry in advancing its shipping service at the cost of employment opportunities for seafarers? Will the introduction of MASS lead to technological unemployment and skill degradation? The answers to these and many other relevant questions are still uncertain and need to be investigated further.

Furthermore, as a result of the review undertaken in this paper, the following recommendations for future research should be given:

Firstly, as noted by Lee and Seppelt (2009, p. 417), "*A common fallacy is that automation can improve system performance by eliminating human variability and errors*." A number of scholars have been very critical towards the statement that automation would eliminate negative implications of human factors (Perrow, 2011). The actual safety benefits of MASS in comparison to conventional vessels need to be further explored. Engineering and design efforts should also aim for closer cooperation with the objective to fill the knowledge gaps among designers, users and human factor experts to enable a user-centered approach in ship design rather than the traditional technology-centered approach.

Secondly, the increased level of automation will gradually change the role of human operators from direct control to more indirect supervisory activities, which could pose many human factors challenges as mentioned in previous sections of this paper. Anticipating the future skillsets and competence requirements for shipboard officers and remote-control operators under every autonomous shipping scenario in order to establish a common framework to guide future maritime training and education efforts would also be an important step forward to MASS operations. Future research could extend the existing studies to obtain a more complete picture regarding both technical and non-technical skills needed for operation of MASS. Furthermore, to ensure a cohesive, well-rounded approach to MASS development, governance and oversight, discussions at IMO should continue with the objective to identify the emerging risks, mitigation strategies and to generate early standards, ethics guidelines, governance principles and best practices to reduce potential safety concerns while harnessing the full potential of the emerging technology.

5 Conclusion

New vessel operational concepts based on remote-control and autonomous vessel functions hold great safety, economic and environmental promise for the maritime industry. Improved optimization of ship operations, and the transition towards autonomous shipping would be a natural progression in the evolution of shipping on the macro level. This paper provided a review of relevant research studies relating to autonomous shipping, explored the challenges associated with MASS operation and concluded with future research suggestions.

To fully realize the potentials of emerging technologies and to unlock the long-term value of MASS, the path forward should be guided by collaboration, cooperation and knowledge sharing among all stakeholders in the maritime industry. Along with technological development, creating an appropriate MASS governance model, a reskilling and upskilling framework, as well as a safety and security management structure to prepare for and to minimize the potentials of any adverse effects that may arise from MASS operations would be essential for the road ahead.

References

Abilio Ramos, M., Utne, I. B., & Mosleh, A. (2019). Collision avoidance on maritime autonomous surface ships: Operators' tasks and human failure events. *Safety Science, 116*, 33–44.

Acanfora, M., Krata, P., Montewka, J., & Kujala, P. (2018). Towards a method for detecting large roll motions suitable for oceangoing ships. *Applied Ocean research, 79*, 49–61.

Ahvenjärvi, S. (2016). The human element and autonomous ships. *TransNav: International Journal on Marine Navigation and Safety of Sea Transportation, 10*, 517–521.

Ait Allal, A., El Amrani, L., Haidine, A., Mansouri, K., & Youssfi, M. (2020). Implementation of 5G communication network for a safe operation of autonomous and conventional ships. *International Journal of Engineering Research in Africa, 51*, 229–248.

Ait Allal, A., Mansouri, K., Mohamed, Y., & Qbadou, M. (2017). *Toward a reliable sea water central cooling system for a safe operation of autonomous ship*. Paper presented at the international conference on recent innovations in engineering and technology (ICRIET) at Berlin, Germany.

Akbar, A., Aasen, A. K. A., Msakni, M. K., Fagerholt, K., Lindstad, E., & Meisel, F. (2020). An economic analysis of introducing autonomous ships in a short-sea liner shipping network. *International Transactions in Operational Research, 28*, 1740–1764.

Allianz. (2020). Safety and shipping review 2019: An annual review of trends and developments in shipping losses and safety. *Allianz Global Corporate & Specialty (AGCS).*

Bačkalov, I. (2020). Safety of autonomous inland vessels: An analysis of regulatory barriers in the present technical standards in Europe. *Safety Science, 128*, 104763.

Bainbridge, L. (1983). Ironies of automation. In *Analysis, design and evaluation of man–machine systems* (pp. 129–135). Elsevier.

Bennett Moses, L. (2011). Agents of change: How the law 'copes' with technological change. *Griffith Law Review, 20*(4), 763–794.

Bitar, G., Breivik, M., & Lekkas, A. M. (2018). Energy-optimized path planning for autonomous ferries. *IFAC-PapersOnLine, 51*(29), 389–394.

Brettel, M., Friederichsen, N., Keller, M., & Rosenberg, M. (2014). How virtualization, decentralization and network building change the manufacturing landscape: An industry 4.0 perspective. *International Journal of Mechanical, Industrial Science and Engineering, 8*(1), 37–44.

Buckley, J. A., Thompson, P. B., & Whyte, K. P. (2017). Collingridge's dilemma and the early ethical assessment of emerging technology: The case of nanotechnology enabled biosensors. *Technology in Society, 48*, 54–63.

Burns, R. (2001). *Advanced control engineering*. Elsevier.

Chaal, M., Banda, O. A. V., Glomsrud, J. A., Basnet, S., Hirdaris, S., & Kujala, P. (2020). A framework to model the STPA hierarchical control structure of an autonomous ship. *Safety Science, 132*, 104939.

Chang, C.-H., Kontovas, C., Yu, Q., & Yang, Z. (2021). Risk assessment of the operations of maritime autonomous surface ships. *Reliability Engineering & System Safety, 207*, 107324.

Chen, P., Huang, Y., Papadimitriou, E., Mou, J., & van Gelder, P. (2020). Global path planning for autonomous ship: A hybrid approach of fast Marching Square and velocity obstacles methods. *Ocean Engineering, 214*, 107793.

Chialastri, A. (2012). *Automation in aviation*. INTECH Open.

Collingridge, D. (1982). *The social control of technology*. St. Martin's Press, 200 p.

Daly, A., Hagendorff, T., Li, H., Mann, M., Marda, V., Wagner, B., ... Witteborn, S. (2019). Artificial intelligence, governance and ethics: Global perspectives. In *The Chinese University of Hong Kong Faculty of law research paper (2019–15)*. https://doi.org/10.2139/ssrn.3414805

Dhami, H., & Grabowski, M. (2011). Technology impacts on safety and decision making over time in marine transportation. *Proceedings of the Institution of Mechanical Engineers, Part O: Journal of Risk and Reliability, 225*(3), 269–292.

Dreyer, L. O., & Oltedal, H. A. (2019). *Safety challenges for maritime autonomous surface ships: A systematic review*. Paper presented at the the third conference on maritime human factors, Haugesund.

EMSA. (2019). *Annual overview of marine casualties and incidents 2019*. European Maritime Safety Agency.

Endsley, M. R. (2018). Automation and situation awareness. In *Automation and human performance* (pp. 163–181). Routledge.

Form, W. (1987). On the degradation of skills. *Annual Review of Sociology, 13*(1), 29–47.

Gasser, U., & Almeida, V. A. F. (2017). A layered model for AI governance. *IEEE Internet Computing, 21*(6), 58–62.

Genus, A., & Stirling, A. (2018). Collingridge and the dilemma of control: Towards responsible and accountable innovation. *Research Policy, 47*(1), 61–69.

Hagemann, R., Huddleston, J., & Thierer, A. (2018). Soft law for hard problems: The governance of emerging technologies in an uncertain future. *Colo Tech LJ, 17*, 37.

Hancock, P. A. (2019). Some pitfalls in the promises of automated and autonomous vehicles. *Ergonomics, 62*(4), 479–495.

Hassani, V., & Lande, S. V. (2018). Path planning for marine vehicles using Bézier curves. *IFAC-PapersOnLine, 51*(29), 305–310.

Heffner, K., & Rødseth, Ø. J. (2019). *Enabling technologies for maritime autonomous surface ships*. Paper presented at the Journal of Physics: Conference series.

Hofmann, E., & Rüsch, M. (2017). Industry 4.0 and the current status as well as future prospects on logistics. *Computers in Industry, 89*, 23–34.

Honekamp, W. (2018). Electronic navigation challenges for autonomous ships. *Mobility in a Globalised World 2017, 19*, 211.

Hoyhtya, M., Huusko, J., Kiviranta, M., Solberg K., & Rokka, J. (2017). *Connectivity for autonomous ships: Architecture, use cases, and research challenges*. Paper presented at the international conference on information and communication technology convergence: ICT Convergence Technologies leading the fourth industrial revolution, ICTC 2017.

Huang, Y., Chen, L., Negenborn, R. R., & van Gelder, P. H. A. J. M. (2020). A ship collision avoidance system for human-machine cooperation during collision avoidance. *Ocean Engineering, 217*, 107913.

Huddleston, J. (2019). *Soft law and emerging technology in the states* (p. 19). The James Madison Institute.

IHS. (2019). Shipping fleet statistics 2019. Department for Transport. https://assets.publishing.service.gov.uk/government/uploads/system/uploads/attachment_data/file/951092/shipping-fleet-statistics-2019.pdf.

IMO. (2018a). *Scoping exercise begins, to look into how safe, secure and environmentally sound Maritime Autonomous Surface Ship operations may be addressed in IMO instruments*. Retrieved January 4, 2020.

IMO. (2018b). Working group report in 100th session of IMO Maritime Safety Committee for the regulatory scoping exercise for the use of maritime autonomous surface ships (MASS). *MARITIME SAFETY COMMITTEE 100th session* MSC 100/WP.8.
IMO. (2011). *International convention on standards of training, certification and watchkeeping for seafarers (STCW) 1978, as amended in 1995/2010*. London: International Maritime Organisation.
Johansen, T. A., Cristofaro, A., & Perez, T. (2016). Ship collision avoidance using scenario-based model predictive control. *IFAC-PapersOnLine, 49*(23), 14–21.
Johansen, T. A., Perez, T., & Cristofaro, A. (2016). Ship collision avoidance and COLREGS compliance using simulation-based control behavior selection with predictive hazard assessment. *IEEE Transactions on Intelligent Transportation Systems, 17*(12), 3407–3422.
Karlis, T. (2018). Maritime law issues related to the operation of unmanned autonomous cargo ships. *WMU Journal of Maritime Affairs, 17*(1), 119–128.
Kavallieratos, G., Diamantopoulou, V., & Katsikas, S. K. (2020). Shipping 4.0: Security requirements for the cyber-enabled ship. *IEEE Transactions on Industrial Informatics, 16*(10), 6617–6625.
Kavallieratos, G., Katsikas, S., & Gkioulos, V. (2018). Cyber-attacks against the autonomous ship. In *Computer security* (pp. 20–36). Springer.
Kim, J., & Kim, S. (2018). Motion control of multiple autonomous ships to approach a target without being detected. *International Journal of Advanced Robotic Systems, 15*(2), 1729881418763184.
Kim, T.-e., & Mallam, S. (2020). A Delphi-AHP study on STCW leadership competence in the age of autonomous maritime operations. *WMU Journal of Maritime Affairs, 19*, 163–181.
Kim, T.-e., Nazir, S., & Øvergård, K. I. (2016). A STAMP-based causal analysis of the Korean Sewol ferry accident. *Safety Science, 83*, 93–101.
Kim, T.-e., Sharma, A., Gausdal, A. H., & Chae, C.-J. (2019). Impact of automation technology on gender parity in maritime industry. *WMU Journal of Maritime Affairs, 18*(4), 579–593.
Komianos, A. (2018). The autonomous shipping era. Operational, regulatory, and quality challenges. *TransNav: International Journal on Marine Navigation and Safety of Sea Transportation, 12*, 335–348.
Kooij, C., Loonstijn, M., Hekkenberg, R., & Visser, K. (2018). Towards autonomous shipping: Operational challenges of unmanned short sea cargo vessels. In *Marine Design XIII* (pp. 871–880). CRC Press.
Kretschmann, L., Burmeister, H. C., & Jahn, C. (2017). Analyzing the economic benefit of unmanned autonomous ships: An exploratory cost-comparison between an autonomous and a conventional bulk carrier. *Research in Transportation Business and Management, 25*, 76–86.
Larrazabal, J. M., & Peñas, M. S. (2016). Intelligent rudder control of an unmanned surface vessel. *Expert Systems with Applications, 55*, 106–117.
Lee, J., & Seppelt, B. (2009). Human factors in automation design. In *Springer handbook of automation* (pp. 417–436). Springer.
Lee, T. K. (2016). *Liability of autonomous ship: The Scandinavian perspective. How the liability regimes shall be regulated in the Scandinavian region?* University of Oslo (UiO) Thesis.
Lee, Y. I., Kim, S. G., & Kim, Y. G. (2015). Fuzzy relational product for collision avoidance of autonomous ships. *Intelligent Automation and Soft Computing, 21*(1), 21–38.
Leveson, N. (2011). *Engineering a safer world: Systems thinking applied to safety*. MIT Press.
Li, S., & Fung, K. S. (2019). Maritime autonomous surface ships (MASS): Implementation and legal issues. *Maritime Business Review, 4*(4), 330–339.
Liebert, W., & Schmidt, J. C. (2010). Collingridge's dilemma and technoscience. *Poiesis & Praxis, 7*(1–2), 55–71.
Litman, T. (2020). *Autonomous vehicle implementation predictions: Implications for transport planning*. Victoria Transport Policy Institute https://www.vtpi.org/avip.pdf.
Lutzhoft, M., Hynnekleiv, A., Earthy, J. V., & Petersen, E. S. (2019). *Human-centred maritime autonomy-An ethnography of the future*. Paper presented at the Journal of Physics: Conference Series.
Lützhöft, M. H., & Dekker, S. W. A. (2002). On your watch: Automation on the bridge. *Journal of Navigation, 55*(1), 83–96.

Lyu, H., & Yin, Y. (2018). Fast path planning for autonomous ships in restricted waters. *Applied Sciences (Switzerland), 8*(12), 2592.

Lyu, H., & Yin, Y. (2019). COLREGS-constrained real-time path planning for autonomous ships using modified artificial potential fields. *Journal of Navigation, 72*(3), 588–608.

Marchant, G. (2011). The growing gap between emerging technologies and the law. In *The growing gap between emerging technologies and legal-ethical oversight* (Vol. Vol. 7, pp. 19–33). Springer.

Marchant, G. E., Allenby, B. R., & Herkert, J. R. (2011). *The growing gap between emerging technologies and legal-ethical oversight: The pacing problem* (Vol. Vol. 7). Springer.

Mayne, A. J. (1982). Reviewed work: The social control of technology by D. Collingridge. *The Journal of the Operational Research Society, 33*(8), 771–772.

Moses, L. B. (2007). Recurring dilemmas: The law's race to keep up with technological change. *U Ill JL Tech & Pol'y, 239*.

MSC85. (2009). Report of the maritime safety committee on its eighty-fifth session. *MSC, 85*, 26.

MSC98/23. (2017). *Report of the Maritime Safety Committee on its ninety-eighth session, International Maritime Organization*. London: Maritime Safety Committee 98th session Agenda item 23.

MSC99/5/6. (2018). *Regulatory scoping exercise for the use of Maritime Autonomous Surface Ships (MASS) considerations on definitions for levels and concepts of autonomy, International Maritime Organization*. London: Maritime Safety Committee 99th session Agenda item 5.

Namgung, H. (2019). Spectrum requirements for control and non-payload communication of maritime autonomous surface ship. *Advances in Science, Technology and Engineering Systems, 4*(5), 294–299.

National Research Council. (2014). *Autonomy research for civil aviation: Toward a new era of flight*. National Academies Press.

OECD. (2020). *Technology governance*. Retrieved January 9, 2020, from http://www.oecd.org/sti/science-technology-innovation-outlook/technology-governance/.

Pazouki, K., Forbes, N., Norman, R. A., & Woodward, M. D. (2018). Investigation on the impact of human-automation interaction in maritime operations. *Ocean Engineering, 153*, 297–304.

Perrow, C. (2011). *Normal accidents: Living with high risk technologies*. Princeton university press.

Porathe, T., Hoem, Å. S., Rødseth, Ø. J., Fjørtoft, K. E., & Johnsen S. O. (2018). At least as safe as manned shipping? Autonomous shipping, safety and "human error". *Safety and reliability–safe societies in a changing world. Proceedings of ESREL 2018, June 17–21, 2018, Trondheim, Norway*.

Pribyl, S. T., & Weigel, A. M. (2018). How an emerging disruptive technology is poised to impact the maritime industry much sooner than anticipated. RAIL robotics. *Robotics, Artificial Intelligence & Law, 1*(1), 17–25.

Raunek. (2020). *10 Things to consider while using auto-pilot system on ships*. Retrieved January 4, 2021.

Reason, J. (1995). Understanding adverse events: Human factors. *Quality in Health Care: QHC, 4* (2), 80–89.

Reddy, S., Allan, S., Coghlan, S., & Cooper, P. (2020). A governance model for the application of AI in health care. *Journal of the American Medical Informatics Association, 27*(3), 491–497.

Rigaud, E., Lützhöft, M., Kircher, A., Schröder-Hinrichs, J.-U., Baldauf, M., Jenvald, J., & Porathe, T. (2012). Impact: More than maritime risk assessment. *Procedia-Social and Behavioral Sciences, 48*, 1848–1854.

Ringbom, H. (2019). Regulating autonomous ships—Concepts, challenges and precedents. *Ocean Development and International Law, 50*(2–3), 141–169.

Roberts, G. N., Sutton, R., Zirilli, A., & Tiano, A. (2003). Intelligent ship autopilots—A historical perspective. *Mechatronics, 13*(10), 1091–1103.

Salathé, M., Wiegand, T., & Wenzel, M. (2018). *Focus group on artificial intelligence for health*. arXiv preprint arXiv:1809.04797.

Schröder-Hinrichs, J.-U., Praetorius, G., Graziano, A., Kataria, A., & Baldauf, M. (2016). *Introducing the concept of resilience into maritime safety*. Paper presented at the 6th symposium on resilience engineering, Lisbon, Portugal, June 22–25, 2015.

Schwab, K. (2016). The fourth industrial revolution: What it means, how to respond. *World Economic Forum*. https://www.weforum.org/agenda/2016/01/the-fourth-industrial-revolution-what-it-means-and-how-to-respond/.

Sharma, A., Kim, T., Nazir, S., & Chae, C. (2019). Catching up with time? Examining the STCW competence framework for autonomous shipping. In *Ergoship conference* (pp. 87–93).

Sharma, A., & Kim, T.-e. (2021). Exploring technical and non-technical competencies of navigators for autonomous shipping. *Maritime Policy & Management*, 1–19.

Shuai, Y., Li, G., Cheng, X., Skulstad, R., Xu, J., Liu, H., & Zhang, H. (2019). An efficient neural-network based approach to automatic ship docking. *Ocean Engineering, 191*, 106514.

Statheros, T., Howells, G., & McDonald-Maier, K. (2008). Autonomous ship collision avoidance navigation concepts, technologies and techniques. *Journal of Navigation, 61*(1), 129–142.

Stilgoe, J. (2018). Machine learning, social learning and the governance of self-driving cars. *Social Studies of Science, 48*(1), 25–56.

Tam, K., & Jones, K. D. (2018). Maritime cybersecurity policy: The scope and impact of evolving technology on international shipping. *Journal of Cyber Policy, 3*(2), 147–164.

UNCLOS. (1994). United Nations convention on the law of the sea (UNCLOS). *United Nations*.

UNCTAD. (2020). *Review of maritime transport 2020* (report). Retrieved from https://unctad.org/webflyer/review-maritime-transport-2020.

Valdez Banda, O. A., Kannos, S., Goerlandt, F., van Gelder, P. H. A. J. M., Bergström, M., & Kujala, P. (2019). A systemic hazard analysis and management process for the concept design phase of an autonomous vessel. *Reliability Engineering and System Safety, 191*, 106584.

Velagic, J., Vukic, Z., & Omerdic, E. (2003). Adaptive fuzzy ship autopilot for track-keeping. *Control Engineering Practice, 11*(4), 433–443.

Ventikos, N. P., Chmurski, A., & Louzis, K. (2020). A systems-based application for autonomous vessels safety: Hazard identification as a function of increasing autonomy levels. *Safety Science, 131*, 104919.

Vojković, G., & Milenković, M. (2020). Autonomous ships and legal authorities of the ship master. *Case Studies on Transport Policy, 8*(2), 333–340.

Volz, K., Yang, E., Dudley, R., Lynch, E., Dropps, M., & Dorneich, M. C. (2016). *An evaluation of cognitive skill degradation in information automation*. Paper presented at the proceedings of the human factors and ergonomics society annual meeting.

Wahlström, M., Hakulinen, J., Karvonen, H., & Lindborg, I. (2015). Human factors challenges in unmanned ship operations–insights from other domains. *Procedia Manufacturing, 3*, 1038–1045.

WMU. (2019a). *Transport 2040: Automation, technology, employment - The future of work*. World Maritime University.

WMU. (2019b). *Transport 2040: Autonomous ships: A new paradigm for Norwegian shipping-Technology and transformation*. World Maritime University.

Wróbel, K., Montewka, J., & Kujala, P. (2017). Towards the assessment of potential impact of unmanned vessels on maritime transportation safety. *Reliability Engineering & System Safety, 165*, 155–169.

Wróbel, K., Montewka, J., & Kujala, P. (2018). Towards the development of a system-theoretic model for safety assessment of autonomous merchant vessels. *Reliability Engineering and System Safety, 178*, 209–224.

Yanchin, I., & Petrov, O. (2020). Towards autonomous shipping: Benefits and challenges in the field of information technology and telecommunication. *TransNav: International Journal on Marine Navigation and Safety of Sea Transportation, 14*(3), 611–619.

Zaccone, R., & Martelli, M. (2020). A collision avoidance algorithm for ship guidance applications. *Journal of Marine Engineering and Technology, 19*(sup1), 62–75.

Zhao, L., & Roh, M. I. (2019). COLREGs-compliant multiship collision avoidance based on deep reinforcement learning. *Ocean Engineering, 191*, 106436.

New Business Models for Shipping: Innovation in the Netherlands

Albert W. Veenstra

Abstract The shipping industry is a playground for innovation in technology and business models. On the basis of the Dutch maritime innovation agenda, I provide an overview of innovations and research in three main areas: zero-emission shipping, digital and autonomous shipping and safe shipping. In all these areas, parties in the Netherlands are conducting research and developing new technologies and solutions. I discuss two cases where start-up companies have emerged out of this R&D effort, that are developing new business models in the shipping industry.

1 Introduction

The shipping industry is often termed conservative.[1] This does not mean that there is no research and innovation, however. This chapter outlines the state of innovation in shipping in the Netherlands.

The Netherlands is a maritime nation, and trade and shipping have been two important pillars of its economy for centuries. Still today, the combined transport and logistics industry is a significant part of total Dutch GDP. The Dutch maritime related industries collectively form the so-called Maritime Cluster.[2] This cluster generates about 3% of total Dutch GDP and, directly and indirectly, creates 285.000 jobs.

The Cluster approach to the maritime sector emphasizes the interaction between the different industrial segments of economic activities: shipping, logistics, business services, shipbuilding, ship suppliers, maritime engineering, fishing, and the navy are all interlinked to a greater of lesser extent. This cohesion helps in keeping maritime knowledge at a state-of-the-art level, allows for interesting maritime career

[1]See, for instance, Glave et al. (2014)

[2]See maritimebyholland.com

A. W. Veenstra (✉)
Rotterdam School of Management, Rotterdam, the Netherlands
e-mail: veenstra@rsm.nl

B.-W. Ko, D.-W. Song (eds.), *New Maritime Business*, WMU Studies in Maritime Affairs 10, https://doi.org/10.1007/978-3-030-78957-2_11

paths, strengthens the export position of Dutch maritime companies, and supports innovation in shipping.

In this chapter, I will provide an overview of the innovation strategy of the Dutch maritime cluster, and explore the main innovation roadmaps. I will also introduce some interesting cases of innovative projects that are developing at this point in time and that present important new business models in shipping.

In the next section, I will first position maritime innovation in the broader innovation philosophy in the Netherlands. I will then zoom in on the innovation missions for shipping in some more detail. In the section after that, I will discuss two typical innovation cases that have been developed in public private partnerships. I finish with some concluding remarks.

2 Innovation in Shipping in the Netherlands

The Netherlands is following the European Union (EU) and the United Nations (UN) in the formulation of a mission-oriented innovation strategy. While the UN has formulated 17 sustainable development goals, the EU has formulated five missions on, for instance, cancer, climate change and water quality. In the Netherlands, we have five Innovation Missions, and a sixth so-called cross cutting theme on Key Enabling Technologies.[3] The five missions are:

- Energy Transition and Sustainability,
- Agriculture, Food and Water,
- Health and Care
- Security
- Industry Earning Capacity.

The Dutch innovation system is structured in such a way that there is a joint agreement on funding the research and development for these missions between governments, knowledge institutes (universities, government research institutes) and business. This is a typical triple-helix approach to innovation (see, for instance, Etzkowitz, 2003).

One further element in this triple helix approach is that there are a 17 special public-private partnerships (PPP) that represent parts of the Dutch economy (energy industry, chemical industry, logistics industry, and so on), who also have specific innovation funding at their disposal. These PPPs are called Top Consortium for Knowledge and Innovation, or TKI. The maritime industry has its own TKI that is

[3]See https://www.topsectoren.nl/missiesvoordetoekomst (in Dutch). For an English summary, https://www.topsectoren.nl/missiesvoordetoekomst/documenten/publicaties/2019-publicaties/september-2019/23-09-19/factsheet-dutch-solutions-to-grand-challenges

part of the water management industry (also a strong Dutch economic activity), in which there are also two TKIs covering delta technology and water management.[4]

Important societal challenges for the shipping industry are a bit scattered in this mission-oriented innovation framework:

1. The transport part of shipping can be found under the transport and mobility innovation agenda, which is part of the mission Energy Transition and Sustainability,
2. Some technical challenges are part of the cross-cutting theme Key Enabling Technologies,
3. The generic maritime innovation agenda is part of the mission Agriculture, Food and Water.

Given this relative fragmentation, the Maritime Cluster has developed a comprehensive innovation agenda. In fact, this agenda builds on earlier innovation agendas that have driven the industry's technological pioneering role of decades.

Recently, this effort of maintaining a healthy and topical innovation agenda for the Dutch maritime industry was newly institutionalised in an innovation platform called the Netherlands Forum Smart Shipping, or Smash![5] The aim of this platform is to bring together all the parties from the public and private sector to specifically advance the agenda towards autonomous shipping. Smash! is the Dutch representative in the International Network of Autonomous Shipping (INAS).[6] This underlines the international nature of innovation in shipping, and of the Dutch maritime cluster.

In the next section, I will discuss the Dutch maritime innovation agenda in some detail.

3 Three Innovation Missions for Shipping

The Dutch maritime cluster has an innovation agenda that consists of three main sections:

1. Towards zero emission shipping,
2. Digital and autonomous shipping,
3. Safety in shipping

Note that these three sections are very much oriented towards the ship: the ship needs to become cleaner, smarter and safer. This is quite a different view on innovation in shipping than, for instance, in Lorange's (2020) book "Innovations in Shipping". Lorange concentrates on innovations in shipping business: paper

[4]For information on TKI Maritiem, see https://www.tkimaritiem.nl/ (in Dutch)

[5]See smashnederland.nl/english. Smash! stands for Smart Shipping: highly automated sailing at sea, in ports and on inland waterways.

[6]http://www.autonomous-ship.org/index.html

trading, innovation management, marketing and shipping finance. We will see, however, that the innovation approach of the Dutch Maritime Cluster also leads to innovation in shipping business, but takes these innovations in a very different direction.

3.1 Towards Zero Emission Shipping

The exhaust emissions, but also the other pollution ships cause in normal operations, have been on the agenda for a long time. The fact that we have the UN convention on the prevention of pollution from ships (MARPOL) since 1973 attests to this. Annex VI, that was added to MARPOL in 1997, deals with atmospheric pollution of ships. This Annex covers a wide range of emissions and has also given rise to the designation of Emission Control Areas (ECA) in which strict regulations apply to sulphur emissions. Limits on sulphur emissions outside the ECA have also been cut, as per 1 January 2020.

At the same time, international shipping (as well as transport by air) has been excluded from the Paris Accords, and thus the shipping industry is still clinging to the relatively unambitious 50% reduction of CO_2 emissions by 2050. It seems, however, that the industry is finding its way with initiatives such as the family of large LNG container vessels of CMA-CGM, with new engine technology and a promise of a 20% reduction of CO_2 emissions. Sulphur, nitrogen and particle matter emissions are all cut by more than 90% for this vessel.

The main ambition of the Dutch agenda for zero emission shipping is the introduction of 30 zero-emission ships by 2030. The innovation masterplan contains plans for the required financing and knowledge development to achieve this goal. The Dutch government, through the Department of Defence and the national agency for infrastructure management, who are both substantial fleetowners, are involved as launching customers. The main source of financing is the Dutch maritime industry itself, but for about 25% of the funding a bid for the European Recovery & Resilience Fund[7] will be developed.

The research agenda for this part of the innovation program consists of the following elements (NML, 2020):

1. System transition towards zero emission maritime industry
2. Methanol power and energy systems
3. Hydrogen power and energy systems
4. Future power and energy design
5. Modular ship design and production
6. Smart digital monitoring and maintenance
7. Sustainable digital autonomous operations

[7] See https://ec.europa.eu/info/business-economy-euro/recovery-coronavirus/recovery-and-resilience-facility_en

Note that this research agenda contains the obvious research into hydrogen as a power source, but also methanol. In fact, in shipping, a lot of fuel alternatives are on the table, such as LNG, but also ammonia (part of the future power research program). In addition, heavy duty propulsion requires additional development of hydrogen technology, and more insight in the benefits of hybrid solutions for manoeuvring, long haul sailing and so on. Dutch hydrodynamics research institute MARIN has therefore developed a special zero emission lab, in which many different power train configurations can be built and tested within their operational context.

The final two elements in the research agenda indicate that zero emission shipping is not just a matter of finding new fuel options and engine technology. Better management and control of the shipping operation (efficiency), cutting down waste in shipbuilding, maintenance and repair, and reduction of water resistance, vibrations, and noise all contribute to the goal of this agenda. More traditional topics such as anti-fouling, wind assisted propulsion, hull designs and propellor noise will also receive attention.

An important part of the search for new types of power and energy systems is the research into new energy carriers and storage devices, as well as in efficient refit solutions. Since for shipping, everything is enlarged, additional challenges rise from the development of large batteries, supercondensators, and heat storage solutions. Currently, the electrification in some inland shipping vessels is limited by the size of the batteries that need to be taken onboard of these vessels, extending their width, or adding substantial weight to the ship (which results in loss of cargo carrying capacity). Refitting is a common way to modernize ships, and this generates important research questions on changes in ship design and operations.

While this description of the shipping innovation agenda in the Netherlands concentrates on ships, some port related innovations will also help achieve the goal of zero-emission shipping. The main idea in this context is shore power in ports. Ships in port still consume a lot of energy through control systems, crew quarters and other auxiliary systems and cargo handling systems. While shore power is already available for ships with modest energy needs, heavy duty shore power requires substantial investment in on-shore installations, the electricity grid and power generation. Large ports in the Netherlands (Rotterdam, Amsterdam) have programs in place for inland barging, ferries and river cruise ships. For ocean shipping, research and development programs are in place to experiment with existing technology, and develop new solutions.

A final element in the agenda on zero emission shipping is the renewed attention for circularity in shipbuilding and shipping. While the shipping industry, due to its substantial ship scrapping activities, is a great example of a classic circular industry, there is also much to criticise. More research into the use of materials, design for re-use and recycling, material separation technology, and circularity in maintenance and operations will support the development of a more environmentally responsible circular economy in shipping.

The zero-emission shipping agenda relies strongly on ship and energy related technology. There is also another path that at the very least supports the development towards zero-emission shipping. This part is the optimization of ship operations: optimal loading of ships and fuel efficient sailing. It is this area that has led to some interesting new start-up companies in the Netherlands.

One of these companies is called We4Sea (we4sea.com) and it offers performance monitoring and emission reporting solutions to shipping companies, ship managers, charterers and ship designers. The business model derives from the inaccuracy of manual reporting by shipping companies on performance and emission monitoring, as compared to sensor-based and, even better, model-based solutions. The company can develop digital-twins of ships in which all relevant information of the ship and sailing conditions can be collected, and used to support operational decision making (We4Sea, undated).

Performance improvement has also sparked another new business venture, which will be discussed in some more detail in Sect. 4.

3.2 *Digital and Autonomous Shipping*

The development of fully autonomous ships is not new. The EU funded MUNIN project, for instance, started in 2012.[8] The aim of the MUNIN project was to develop a concept of an autonomous ship, which was guided by onboard autonomous systems, but with remote control onshore.

In the meantime, the world has moved on, and autonomous shipping is not fiction anymore. In fact, in the Netherlands alone, there are a number of examples with drones cleaning rivers, or taking depth or water quality measurements, and in late Summer 2019, an experiment started with an autonomous inland barge of 135 m LOA. This test lasted until May 2020. Another experiment at sea, with collision avoidance technology installed in an offshore service vessel was conducted in March 2019. This experiment was part of a large scale Dutch joint industry project called Autonomous Shipping (2017–2019).

The promise of autonomous shipping is the reduction of costs, such as crew, reduced energy use as a result of more efficient operations, safer operations as a result of the elimination of human error, and increased transport efficiency as a result of better integration with other modes of transport and logistics.

This promise already points at a more complex problem than just developing an autonomous ship. Automation and digitisation go hand in hand, because the entire system in which the ship operates needs to be ready for automation: the waterway infrastructure, ports, service providers, traffic management. The Dutch innovation agenda for autonomous shipping specifically includes the transport function of shipping, and foresees new business models for the operation of ships, along the lines of disruptive businesses such as Über.

The main elements of the research agenda are:

[8]See http://www.unmanned-ship.org/munin/

1. Business models
2. Safe operations
3. Technology and standardisation
4. Rules & regulations
5. Human factors

The work on business models focuses primarily on the potential cost reductions of autonomous technology, and the balance between costs and benefits of automation, as well as the proliferation effect of introducing autonomous technology on the rest of the operational chain.

Part of the research within safe operations goes into the development of simulation environments in which the safety of operations and traffic management can be assessed in a controlled environment. This requires the formulation of a set of safety criteria, as well as new solutions for safety issues in shipping. Through demonstrators, a safety assessment approach needs to be developed and validated that can be used to certify ships and traffic management systems in real-life.

The actual technical solution for a fully automated ship is highly complex, and still requires a lot of research and innovation. An overview of a schematic of the architecture of an autonomous ship is presented in Fig. 1 below.

This means that there is much work to do in developing individual components of the system architecture for autonomous ships, and that the standards required for the interaction of all these components also needs to be formulated. Here, Dutch parties are reaching out to international partners to widen the adoption of these standards. The same effort has to be spent on the systems that connect the ship to its operational environment. Also here, system components and standards need to be developed and broad international adoption achieved.

In the area of rules and regulations, many problems arise, but perhaps one of the most pressing is the issue of liability. Without humans visibly present and in control of operation of ships (and also infrastructure such as bridges and locks, and traffic management), liability becomes a problem that urgently needs workable solutions that do not, at the same time, kill all innovation. The current development of the laws and regulations in this field are very much focused on automated driving cars, and are not yet settled. For shipping, discussions are at the earliest of stages.

Finally, the human factor is a topic in any discussion on the introduction of autonomous technology in human working environment. There is a developing strand of literature that looks into exactly this problem. See for instance Berkers et al. (2019). The research will focus on human-machine interaction, monitor this interaction during experiments and eventually develop training programs for personnel in traffic management positions.

In the public private partnership Smash!, government partners have opened a platform to formally apply for autonomous experiments on the Dutch waterways.[9] To guide these experiments, a policy guideline was developed in the Netherlands to

[9]See https://www.rijkswaterstaat.nl/water/wetten-regels-en-vergunningen/scheepvaart/scheepvaartverkeerswet/smart-shipping/index.aspx

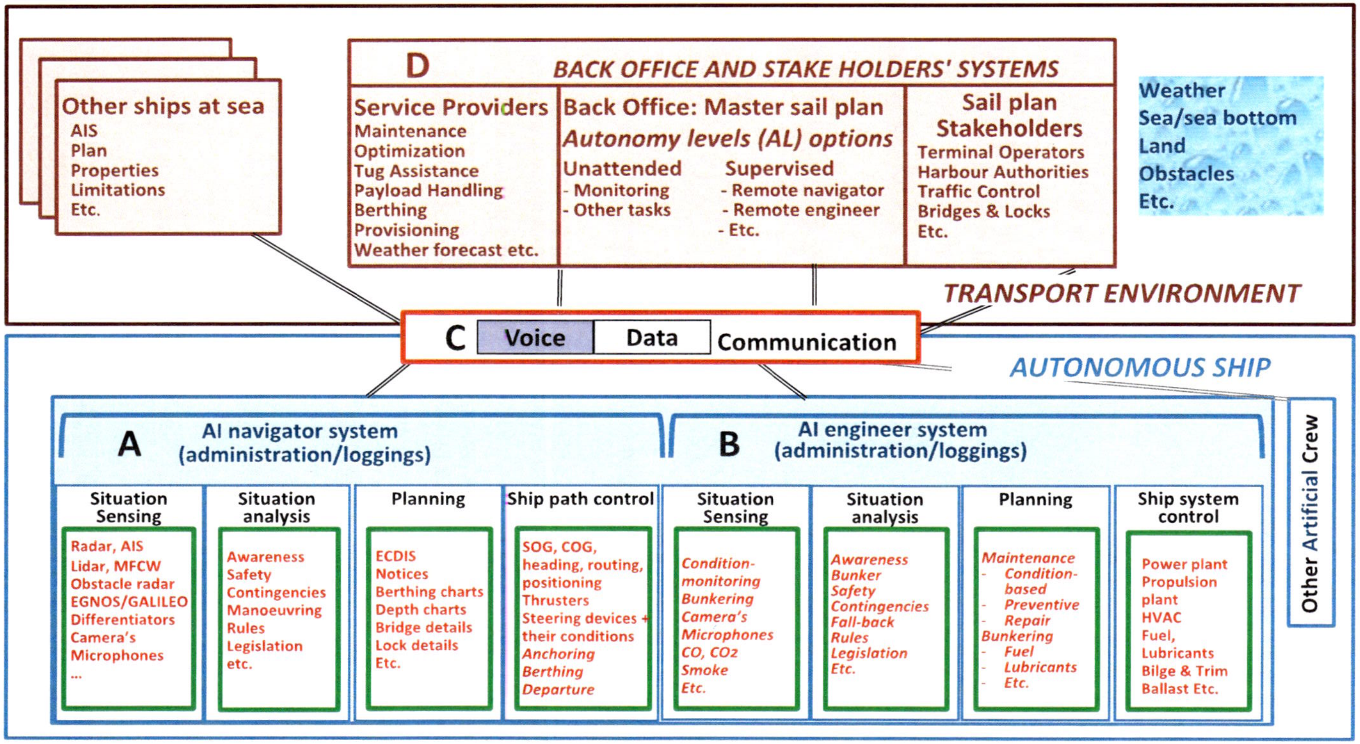

Fig. 1 Autonomous shipping overall model. Sovereign proposal. Source: JIP proposal Autonomous Shipping

provide the formal rules for such experiments, and to allow the Dutch government body for the waterways to evaluate and approve these experiments. This enables innovation in so-called "far reaching autonomous sailing on national waterways".

3.3 *Safety in Shipping*

The view on safety in shipping in the Netherlands is broad, and contains topics from collisions of ships at sea to the loss of cargo and hazardous pollution. The Netherlands is at a crossroads of shipping routes, and also has some busy ports. In addition, a large part of the North Sea area is designated for the development of extensive wind farms, which limits the sailing routes for ships even more. Fairly recent events, such as the loss of more than 300 containers from the MSC Zoe in the sea north of the Netherlands has strengthened this broad view.

There is a long history and track record on research and investigation of the behaviour of ships in sea conditions that drives this safety mission. Knowledge institutes such as MARIN have a world class reputation on stability, manoeuvrability and seakeeping research that forms the basis for the agenda to make shipping safer.

An addition element in this is the safety of cargo onboard and on deck. Lashing, cargo and weight distribution, and the investigation of the related dynamics is also part of this research program.

Further elements in this mission are: collision risk assessment, human factors and virtual reality, safety of autonomous ships and on board support systems. Research in each of these elements build on the expertise in model and simulation-based research, combined with the introduction of new technology such as virtual reality or digital twins. Real life experiments are always the end stage of the innovation pipeline, as was discussed in the automation agenda.

In the remainder, I will discuss two innovation cases that show the type of development of new concepts and business models that are typical for the Dutch maritime innovation agenda.

4 Development of New Business Models

In this section, I will first discuss the case of CoVadem. This is a new company that runs a platform for the cooperative collection of data on inland waterways. As second case, I discuss the line for research on maintenance and service logistics for large maritime assets that has been running for over a decade.

4.1 Business Case: CoVadem

CoVadem stands for cooperative sailing draught measurement (cooperatieve vaardiepte-meting, in Dutch). The narrative here is that rivers, water heights and draughts and riverbeds are in constant movement. Inland shipping lines who travel along rivers must take these changes into account when they take cargo onboard. Their estimates of, for instance, draughts further up or down the river in a few days' time limit their maximum loading capacity now. Shipowners often estimate the loading capacity conservatively to avoid grounding or excessive fuel consumption in shallow sections of the river.

The CoVadem solution is to equip a substantial number of ships with dedicated measurement equipment to collect data on the profile of the riverbed and real-time water levels along the entire river. The data is collected in a platform in order to develop prognoses and performance support tools that can be read directly in onboard systems such as ECDIS waterway maps. For trip preparation, Covadem also offers a dashboard and offers services to support vessel owners in terms of analysing and interpreting their performance data.

The technology for this initiative was developed through a combination an inland shipping support projects, one of them initiated by the Dutch Ministry of Transport around 2014/2015. This technology is a sensor and communication box that communicates via a 4G internet connection to a cloud platform. This box is mounted under the vessel (Fig. 2).

Fig. 2 CoVadem sensor box and tools. Source: CoVade

The roll out of boxes on ships was then further supported by various other innovation funding initiatives in combination with co-financing by the Dutch inland shipping industry. At the point of writing, there are more than 200 vessels actively measuring water depth, and first commercial services are actively available for ship owners and other stakeholders.

The detailed river information from the Covadem platform is also of interest for government agencies and contractors, such as dredging companies. The national agency that maintains the waterways obtains more insight in the utilisation and the morphologic development of the rivers.

A next step in this collective measurement approach is to extend the data collection to also generate collective and individual data on emissions. The measurement of emissions of inland waterway vessels is complex, because the engine specs are inaccurate, the necessary sensors are either not available on board, or do not work properly, and tuning of the engine makes a significant difference in emissions. On top of this, sustainability of all modes of transport receives more and more attention of national and supra-national policymakers. The collective approach that CoVadem has piloted for river depth could be an effective and relatively cheap way of developing a proper emission measurement approach for inland shipping in Europe.

Members of the original CoVadem consortium have started a company to bring the initiative further. The business model for this company is, apart from development funding for new services, revenue from the installation of the box in ships, and subscriptions for the dashboard and map services. Shipping companies that contribute to data collection maintain ownership of their own data, and offer a right to use anonymised data to CoVadem. For this they receive a discount on the fee for the data services.

4.2 *Business Case: Maintenance and Service Logistics*

Ships and other floating assets are capital intensive, and their operation often requires 100% uptime. This means that maintenance is at the same time crucial and, if unplanned, disruptive. Research groups, together with Dutch maritime industry, have developed a series of projects in which the optimization of maintenance and service logistics received a central position.

The core concept of innovation in maintenance and service logistics is the notion that physical products can be combined with services, support and knowledge to better serve the customer, but also to extract greater revenues. See Vandermerwe and Rada (1988) for an early source on this development.

A crucial condition for this development of servitization is the quality of information on the basis of which informed decisions can be made about the planning of maintenance activities and the replacement of components and parts. This is more aptly termed *condition based monitoring*. If this quality is high, business models can shift from the mere selling of products to offering service contracts or even offering service guarantees.

A maritime company that has developed these more advanced business models is the Finnish ship engine manufacturer Wärtsilä. This company offers a range of services related to ship engines from spare part delivery around the world to performance guarantees. The latter includes dedicated technical support on site and remote in order to guarantee the operational reliability and uptime.

A ship engine is only one component, however. Many of the ships in the Netherlands are complex vessels, such as offshore supply vessels, dredging vessels, oil exploration vessels, or navy vessels. Maintenance and service logistics innovations have therefore targeted many of these more complex ship types.

This innovation area also builds on another development that is perhaps already a few decades old: the development towards a circular economy (see, for instance, Dekker et al., 2013). The notion that spare parts or products should not be thrown away as trash, but might be reused, refurbished or recycled was developed decades ago. This way of thinking in supply chains that are cyclical instead of linear has also found its way into the maintenance and service logistics world. Both the idea that refurbished, as good as new, parts can offer cost reductions, as well as the reduction of waste along the entire operational lifecycle of an asset result in interesting new applications and innovations.

This servitization challenge is summarized in the following graphic (Fig. 3).

The main challenge in advanced maintenance and service logistics is to work towards real predictive maintenance processes, instead of the more generic, and inaccurate, planned and reactive maintenance planning. This requires data collection and processing, accurate failure models, and an approach to transform predicted failures in an effective maintenance logistics planning.

Fig. 3 Circular servitization control tower. Source: TKI Dinalog

We will discuss a few examples below where steps towards this goal of integrated maintenance and logistics planning have been attempted, and either failed or succeeded.

First of all, extensive failure models were and are being developed. These have relevance outside the maritime domain as well. Initiatives include all manufacturers of capital intensive assets, such as the railways, lithography printing machines, airplanes, medical equipment. MARIN monitors and predicts remaining fatigue life for many offshore structures at sea.

Research indicated that a model-based approach, in contrast to an experience-based approach, can work well.[10] These models are based on the determination of failure mechanisms (wear, fatigue, corrosion), a numerical approach that connects the life of a component to use or load, and the actual measurement of these inputs, such as use or load from the asset. These models work well in a simulated environment. In many cases, real-life testing is difficult because on many maritime assets, data collection is suboptimal.

The second challenge is to connect condition-based maintenance decisions and logistics planning are integrated. Eruguz et al. (2018) show that potential cost savings of the actual integration of the maintenance and spare part delivery planning can save significant costs. They show this for the case of a cooling fan in a Dutch ocean survey vessel. Cost savings achieved from more integrated planning are about 30%. In the Netherlands, new companies, such as Gordian Logistics Experts, are developing these kinds of advanced services through maintenance and logistics control towers.

Another, related development is the use of new technologies to create spare parts. Additive manufacturing, or 3D printing, is such a technology. In a project on the application of additive manufacturing technologies, a number of case studies were developed, that have led to, among others, a roadmap for additive manufacturing in maintenance and service logistics. A main element in this roadmap is a tool that can be used to select suitable parts and components for additive manufacturing (see Knofius et al., 2016).

Many companies that employ or manufacture capital intensive assets are usually faced with bills of material that contains thousands of parts and components. Not all of these can be 3D-printed, because they contain different materials, or are awkwardly designed for a 3D printer. Based on detailed part and component information that is usually maintained in a material or enterprice resource planning system, the tool ranks parts on various criteria and offers a selection of the most suitable parts. A Dutch company called Dimanex has developed this tool into a working business approach.

---

[10]See WP 1 of the TKI Dinalog project MaselmaBridge.

5 Concluding Remarks

The Dutch maritime industry, collected in a so-called Cluster, has maintained its strong international position by continuous research and innovation. This innovation supports the industry to stay societally relevant: zero-emission, efficient, safe.

Along the lines of the main societal challenges—green, smart, safe—, the Dutch maritime cluster has developed innovation agendas that are ambitious and that play to the strengths of the Dutch maritime knowledge base. Important knowledge institutes such as MARIN, TU Delft, TU Eindhoven, University Twente, but also the Rotterdam School of Management are involved in projects together with the industry.

Apart from successful public private partnerships, the research and innovation also creates new business opportunities. I discussed a number of these, from companies that offer innovative ways of collecting data and supporting operational performance to companies offering complex tools for spare part manufacturing or control towers for advanced maintenance logistics planning. With these companies the Dutch innovation policy achieves its ultimate goal of broadening the earning capacity of the Dutch economy.

Finally, in a successful triple helix model, the government also had to play its role. The Dutch government is doing that by funding research and innovation, through its societal innovation mission approach. It is, however, also contributing by creating the right environment for experimentation and real-life trials. The arrangement to experiment with autonomous ships in public waterways is an example of the innovative mindset of the Dutch government.

We face many societal challenges, not only in shipping. Innovation will help us tackle many of these challenges. In the Netherlands, innovation in a triple helix setting is an important part of the solution.

References

Berkers, H. A., Rispens, S., & Le Blanc, P. M. (2019). *How robots are changing work re-design*. Presented at the EAWOP small group meeting: Antecendents of work design, Amsterdam, the Netherlands, 3–4 June 2019.

Dekker, R., Fleischmann, M., Inderfuhrt, K., & van Wassenhove, L. N. (Eds.). (2013). *Reverse logistics: Quantitative models for closed-loop supply chains*. Springer.

Eruguz, A. S., Tan, T., & van Houtum, G. J. (2018). Integrated maintenance and spare part optimization for moving assets. *IISE Transactions, 50*(3), 230–245.

Etzkowitz, H. (2003). Innovation in innovation: The triple helix of university-industry-government relations. *Social Science Information, 42*(3), 293–337.

Glave, T., Joerss, M., & Saxon, S. (2014). The hidden opportunity in container shipping. *McKinsey & Company Insights*. https://www.mckinsey.com/business-functions/strategy-and-corporate-finance/our-insights/the-hidden-opportunity-in-container-shipping

Knofius, N., Van der Heijden, M. C., & Zijm, W. H. (2016). Selecting parts for additive manufacturing in service logistics. *Journal of Manufacturing Technology Management, 27*(7), 915–931.

Lorange, P. (2020). *Innovation in shipping*. Cambridge University Press.
NML (Nederland Maritime Land). (2020). *Masterplan for an zero-emission maritime industry*. Whitepaper.
Vandermerwe, S., & Rada, J. (1988). Servitization of business: Adding value by adding services. *European Management Journal, 6*(4), 314–324.
We4sea (undated) *The ultimate guide to Model-Based Performance Monitoring systems for ships*. Whitepaper. https://www.we4sea.com/request-whitepaper/